D0220975

The Economics of Landscape and Wildlife Conservation

Edited by

Stephan Dabbert
University of Hohenheim
Germany

Alex Dubgaard
The Royal Veterinary and Agricultural University
Denmark

Louis Slangen
Wageningen Agricultural University
The Netherlands

and

Martin Whitby
University of Newcastle upon Tyne
UK

CAB INTERNATIONAL

CAB INTERNATIONAL
Wallingford
Oxon OX10 8DE
UK

Tel: +44 (0)1491 832111
Fax: +44 (0)1491 833508
E-mail: cabi@cabi.org

CAB INTERNATIONAL
198 Madison Avenue
New York, NY 10016-4314
USA

Tel: +1 212 726 6490
Fax: +1 212 686 7993
E-mail: cabi-nao@cabi.org

A catalogue record for this book is available from the British Library, London, UK.

Library of Congress Cataloging-in-Publication Data
The economics of landscape and wildlife conservation / edited by Stephan Dabbert.
p. cm.
Papers presented at a workshop held at the University of Hohenheim in Sept 1996.
Includes bibliographical references (p. 259) and index.
ISBN 0851992226 (alk. paper)
1. Landscape protection—Economic aspects—Congresses. 2. Agricultural ecology—Economic aspects—Congresses. 3. Wildlife conservation—Economic aspects—Congresses. 4. Landscape protection—European Union countries—Economic aspects—Congresses. 5. Agricultural ecology—Economic aspects—European Union countries—Congresses. 6. Wildlife conservation—Economic aspects—European Union countries—Congresses. I. Dabbert, Stephan, 1958—.
QH75.A1E29 1998
338.4'333372—dc21 97-34941
CIP

ISBN 0 85199 222 6

Typeset in Photina by AMA Graphics Ltd, UK
Printed and bound in the UK by Biddles Ltd, Guildford and King's Lynn

Contents

Contributors

Patrick Baudoux is a Research Assistant in the Department of Farm Management, University of Hohenheim, D70593 Stuttgart, Germany.

Peter Billing is Assistant Administrator at the European Commission, DG XI (Environment), Unit D.1 (water protection, soil conservation, agriculture), Rue de la Loi 200, B-1049 Bruxelles, Belgium.

François Bonnieux is Director of Research and Chairman of the Rural and Environmental Economic Unit, INRA, 65 rue de St Brieuc, F-35042 Rennes, France.

Stephan Dabbert is Professor of Agricultural and Resource Economics at the University of Hohenheim, D-70593 Stuttgart, Germany.

Edi Defrancesco is a Professor of Agricultural Economics in the Department TeSAF, University of Padua, Agricultural Faculty, Agropolis 35020 Legnaro (PD), Italy.

Alex Dubgaard is a Senior Lecturer in the Department of Economics and Natural Resources, Royal Veterinary and Agricultural University, Rolighedsvej 26, 1958 Fredriksborg C (Copenhagen), Denmark.

Alberto Garrido is an Assistant Professor in the Departamento de Economia y Ciencias Sociales Agrarias, Universidad Politécnica de Madrid, ETS Ingenieros Agrónomos, 28040 Madrid, Spain.

Michael Glemnitz is a scientist at the Institute of Land Use Systems and Landscape Ecology at the Centre for Agricultural Landscape and Land Use Research (ZALF), 15374 Müncheberg, Germany.

Gottfried Häring is a Research Assistant in the Department of Farm Management, University of Hohenheim, D70593 Stuttgart, Germany.

Knut Per Hasund is Assistant Professor in Natural Resource and Environmental Economics at the Swedish University of Agricultural Sciences, PO Box 7013, S-75007 Uppsala, Sweden.

Fredrik Holstein is a PhD student in the Department of Economics at the Swedish University of Agricultural Sciences, S-75007 Uppsala, Sweden.

Eva Iglesias is an Assistant Professor in the Departamento de Economia y Ciencias Sociales Agrarias, Universidad Politécnica de Madrid, ETS Ingenieros Agrónomos, 28040 Madrid, Spain.

Clifford Jurgens is an Associate Professor in the Department of Physical Planning and Rural Development, Wageningen Agricultural University, NL-6703 BJ Wageningen, The Netherlands.

Harald Kächele is a scientist at the Institute of Socioeconomics at the Centre for Agricultural Landscape and Land Use Research (ZALF), 15374 Müncheberg, Germany.

Gottfried Kazenwadel is a Research Assistant in the Department of Farm Management, University of Hohenheim, D70593 Stuttgart, Germany.

Uwe Latacz-Lohmann is a lecturer in the Department of Agricultural Economics at Wye College, University of London, Wye, Ashford, Kent TN25 5AH, UK.

Philippe Le Goffe is a senior lecturer in the School of Agriculture (ENSA) and Associate Fellow of INRA, 65 rue de St Brieuc, F-35042 Rennes, France.

Maurizio Merlo is a Professor of Forestry and Environmental Economics in the Department TeSAF, University of Padua, Agricultural Faculty, Agropolis 35020 Legnaro (PD), Italy.

Wilfried Mirschel is a scientist at the Institute of Landscape Modelling at the Centre for Agricultural Landscape and Land Use Research (ZALF), 15374 Müncheberg, Germany.

Andrew Moxey is a lecturer in Agricultural Economics and Food Marketing, University of Newcastle upon Tyne, Newcastle upon Tyne, NE1 7RU, UK.

Arie Oskam is a Professor in Agricultural Economics and Policy, Agricultural University, Hollandseweg 1, 6708 Wageningen, The Netherlands.

Giorgio Osti is a Researcher in the Department of Human Sciences, University of Trieste, Piazzale Europa 1, I-34127 Trieste, Italy.

Hans-Peter Piorr is a scientist at the Department for Landscape Development at the Centre for Agricultural Landscape and Land Use Research (ZALF), D15374 Müncheberg, Germany.

Christopher Ray is a Research Fellow in the Centre for Rural Economy, University of Newcastle upon Tyne, Newcastle upon Tyne, NE1 7RU, UK.

Caroline Saunders is a Senior Lecturer in Economics at Lincoln University, New Zealand.

Louis Slangen is Associate Professor in the Department of Agricultural Economics and Policy, Agricultural University, Hollandseweg 1, 6708 Wageningen, The Netherlands.

Ulrich Stachow is a Research Scientist at the Institute of Land Use Systems and Landscape Ecology at the Centre for Agricultural Landscape and Land Use Research (ZALF), 15374 Müncheberg, Germany.

José Sumpsi is a Professor in the Departamento de Economia y Ciencias Sociales Agrarias, Universidad Politécnica de Madrid, ETS Ingenieros Agrónomos, 28040 Madrid, Spain.

Bareld van der Ploeg is a researcher at the DLO Winand Staring Centre, 6700 AC Wageningen, The Netherlands, and at the DLO Agricultural Research Institute, Den Haag, The Netherlands.

Jaap van Wenum is a PhD researcher in the Department of Farm Management, Wageningen Agricultural University, Hollandseweg 1, NL-6706 KN Wageningen, The Netherlands.

Robert D. Weaver is Professor of Agricultural Economics at the Pennsylvania State University, 207 Armsby Building, University Park, PA 16802, USA.

Martin Whitby is Professor of Countryside Management at the Centre for Rural Economy, University of Newcastle upon Tyne, Newcastle upon Tyne NE1 7RU, UK.

Ben White is a lecturer in Agricultural Economics at the University of Newcastle upon Tyne, Newcastle upon Tyne NE1 7RU, UK.

Julie Whittaker is a lecturer in the Agricultural Economics Unit, School of Business and Economics, University of Exeter, St German's Road, Exeter EX4 6TL, UK.

Ada Wossink is an Associate Professor in the Department of Farm Management, Wageningen Agricultural University, Hollandseweg 1, NL-6706 KN Wageningen, The Netherlands.

Preface

This book arises from a European Union (EU) Concerted Action on Policy Measures to Control Environmental Impacts from Agriculture, under the Agro Industrial Research Programme (AIR3-CT93-1164). The book results from a workshop held at the University of Hohenheim, in September 1996, which brought together academics and policy-makers to examine the critical issues of nature conservation and landscape.

There have been EU policies aimed at conserving wildlife and landscape in Europe for more than a decade, and appraisal of the success of these policies is opportune as the next round of world trade discussions gets under way. In addition to EU policies, there is, within members of the union, a rich and diverse wealth of experience in operating and managing policy mechanisms that have been developed over a longer period.

The chapters which follow reflect that diversity and show the breadth of experience which is now available. They represent a selection of the papers presented at the workshop in Hohenheim, chosen on the basis of peer review by at least two academics. We also acknowledge that it will require more than one volume such as this to make the definitive statement about such activities, but hope that this and the other volumes to be produced from the Concerted Action will make a substantial step in that direction.

The editors are grateful to the EU for funding the workshop and the organization under the Agro Industrial Research Programme (AIR3-CT93-1164), and also to its participants for their contribution. In particular, we thank Arie Oskam of the University of Wageningen for providing the project management, Johannes Umstätter for organizing the workshop and the University of Hohenheim for hosting it. We have also benefited from the editorial advice of Andrew

Moxey, Ben White and Robert Weaver (all contributors to this volume) and from the editorial effort of Richard Hill in assembling the final text for publication and preparing the index.

Stephan Dabbert, Alex Dubgaard, Louis Slangen and Martin Whitby

Towards Sustainable Agriculture: the Perspectives of the Common Agricultural Policy in the European Union

1

Peter Billing[1]

INTRODUCTION

On the basis of a short sketch of the main factors influencing the reform of the Common Agricultural Policy (CAP) in 1992 and some of its major instruments to promote sustainable development of agriculture in Europe, this chapter draws up an interim balance sheet of the 1992 CAP reform, especially with regard to the so-called agri-environmental measures. The analysis focuses on Council Regulation (European Economic Community (EEC)) 2078/92 of 30 June 1992 on agricultural production methods compatible with the requirements of the protection of the environment and the maintenance of the countryside. Subsequently, some results, experiences and problems with this instrument are discussed.

A discourse about the perspectives for a sustainable and environmentally friendly development of European agriculture concludes the chapter. Special attention is drawn to the Commission of the European Communities' *Agricultural Strategy Paper* of December 1995. This strategy paper outlines the general objectives of the future CAP in the light of the envisaged enlargement of the European Union (EU) to the east: increase of international competitiveness, development of an integrated concept for rural areas, and simplification and more regional flexibility for the implementation of European Community legislation. The paper suggests that an expansion and consolidation of the 1992 reform is to be expected. This means that the importance of environmental considerations within CAP measures will further increase.

Agriculture, together with forestry, occupies more than 80% of the area of the European Community. Agriculture has shaped the European landscape and continues to do so. Just as in the case of manufacturing industry and transport, agriculture has undergone significant structural changes in the past 40 years. Among the factors which have brought change are increased mechanization, better transport possibilities and improvements in seed quality, crop protection and animal varieties, as well as international trade and competition in food and foodstuffs. This structural change did not always have a favourable effect on the rural areas. The continuing depopulation of the rural areas highlights this development. From the point of view of the environment, CAP brought about a number of negative consequences, such as nitrate in drinking-water, pesticide residues in food and threats to biodiversity. Consequently, the establishment of a sustainable and environmentally friendly agriculture is one of the priorities in the Community's 1992 Fifth Environmental Action Programme (5th EAP). The year 1992 also marked a turning-point for CAP: with Regulations 2078/92 and 2080/92, the Community created instruments to promote agricultural production methods compatible with the requirements of the protection of the environment and the maintenance of the countryside.

After 5 years, it is time now to draw up an interim balance sheet of the 1992 CAP reform. The following remarks will discuss possible perspectives of CAP on the basis of a description and provisional evaluation of the main agri-environmental instruments and an analysis of the present political framework.

STARTING POSITION

The system of guaranteed prices and export subsidies for agricultural products led to enormous surplus production and environmental damage. Consequently, the CAP was increasingly confronted with heavy criticism. Environmental considerations, however, were not sufficient to initiate the CAP reform of 1992. Rather, three developments were the main driving forces:

- Increasing financial burdens caused by CAP expenditure.
- Unprecedented international criticism of CAP export subsidies, especially in the General Agreement on Tariffs and Trade (GATT) framework and the need to conclude the Uruguay Round negotiations.
- Enlargement of the Community with its associated costs put further pressure on the budget.

Basically, the 1992 CAP reform consists of the following elements: reduction of the guaranteed prices by as much as 30% for some products in connection with direct payments to farmers as compensation for income foregone; compulsory set-aside linked with direct payments; adoption of Regulations 2078/92 and 2080/92, which provided incentives for environmentally friendly agricultural production methods and for afforestation (agri-environmental measures).

Apart from the reduction of surplus production, stabilization of rural incomes and defusing the GATT/World Trade Organization (WTO) problems, promotion of environmentally friendly production methods was one of the aims of the reform. With regard to environmental aspects, Regulations 2078/92 and 2080/92 were designed to help with the implementation of the 5th EAP. The aim of this programme is the integration of environmental protection into other Community policies, according to the provisions of article 130r of the Maastricht Treaty, which stipulates: 'Environmental protection requirements must be integrated into the definition and implementation of other Community policies.'

THE EUROPEAN UNION AGRI-ENVIRONMENTAL MEASURES

The programme package based on Regulation 2078/92 has been the most prominent approach at EU level so far to integrating environmental aspects into agricultural policy. The basis of this innovative approach is the contractual agreement between the state and individual farmers, who receive premiums for certain 'environmental services'.

The design of the individual programmes is left to the member states (MS) or administrative units on a regional level. This corresponds to the subsidiarity principle and ensures, to a certain degree, that programmes are adapted to local needs. The Regulation as such can be regarded as a kind of framework of general requirements. Thus, the main responsibility for programme design and implementation lies with the MS. The programmes are notified to the Commission, which, through a sometimes lengthy process of interservice consultation, recommends adoption or modification of programme proposals. These proposals are then put to the vote of MS representatives in the Agricultural Structure and Rural Development Committee (STAR Committee). The role of the Community is mainly to analyse (which may include suggestions for modifications) and cofinance programmes.

Regulation 2078/92 on the one hand allows for general measures that can be used on the entire ultilized agricultural area. These measures mainly aim at a reduction of fertilizer and pesticide use, the promotion of organic farming and a reduction of livestock density (extensification). The general guideline is no more than 2 livestock units (LU) ha^{-1} (1.4 in sensitive areas), which roughly corresponds to the limit value of 170 kg nitrogen (N) ha^{-1} of the Nitrates Directive.

Regulation 2078/92 further allows for measures that are more specifically targeted at the local environmental situation (care for ecologically valuable biotopes or landscape features, such as hedgerows, river-bank strips or 20-year set-aside for environmental reasons). Finally, the Regulation supports the breeding of rare domestic animal species and promotes training measures for farmers. By 31 December 1995, a total of 115 programmes had been approved for Community part-funding. In total, 4.3 billion European currency units (ECU) are made available for the EU-15 during the first 5-year contract period of Regulation 2078 programmes. In addition, MS have to provide additional national funds, because the programmes are cofinanced.

Some other agri-environmental measures should be mentioned here as well. Regulation 2080/92, for example, promotes forestry in the Community. Its main environmental objectives are to contribute towards forms of countryside management more compatible with environmental balance and to combat the greenhouse effect by absorbing carbon dioxide. The Commission approved national programmes under Regulation 2080/92 that will result in the creation of 650,000 ha of new forests. Cofinancing of the EU amounts to 1.2 billion ECU.

Organic farming is also becoming increasingly important. In some MS (for example, Austria), the land converted to organic farming amounts to 8% already (Lampkin, 1996). Regulation 2092/91 is an important instrument to set Community-wide standards for organic farming.

Apart from the measures described above, some other instruments for the integration of environmental protection into agricultural policy should be mentioned briefly:

- Commission proposals for the integration of environmental aspects into some Community market regimes (for example, green cover after grubbing up vineyards).
- Environmental impact assessments for major projects funded by the Community regional and structural funds. About 11% of the structural funds are made available for environmental protection projects (for example, energy-saving measures). Objective 5b regions ('endangered rural regions') will receive 720 million ECU for environmental measures between 1994 and 1999 (European Commission, 1996).
- The Nitrates Directive (91/676/EEC) of 12 December 1991, requiring MS to establish codes of good agricultural practice and action plans to reduce nitrate leaching from agricultural sources.
- The Community initiative LEADER (Liaison Entre Actions de Développement de l'Economie Rurale) supports innovative approaches for sustainable regional development.

RESULTS AND EXPERIENCES WITH THE AGRI-ENVIRONMENTAL MEASURES

Although a final assessment of the environmental consequences of the 1992 reform and its agri-environmental measures is not yet possible, for several reasons (such as a lack of sufficient data, slow response to measures, long pay-off time of measures introduced), some preliminary lessons can be drawn, especially with regard to Regulation 2078/92.

The balance sheet drawn up in the 1996 progress report of the Commission on the implementation of the 5th EAP is mixed. On the positive side, it states that the total consumption of fertilizer and pesticides decreased by 5–10% per year, in individual regions it even dropped by 40%. For Baden-Württemberg, just to pick out one example, Zeddies and Doluschitz (1996) claim that the regional Regulation 2078 programme 'Marktentlastungs- und Kulturlandschaftsausgleich' (MEKA) and the other measures of the 1992 reform resulted in a reduction of the nitrate quantity by 6450 t. Soil erosion in exposed locations was said to have been reduced by 50–80%. The use of pesticides was reduced by at least 191,200 kg. This suggests favourable long-term consequences for the environment and especially for water quality, although it is hard to distinguish whether these improvements can be attributed mainly to the success of the agri-environmental measures or to the impacts of the reform in general.

In order to get a clearer picture of the environmental effects of the agri-environmental measures, further efforts in the area of measurement, development of indicators and methodologies for evaluation seem necessary. Existing experience suggests that further improvements of the agri-environmental measures are necessary. The following examples demonstrate this.

Premiums paid for environmental schemes have to compete with premiums given to conventional farming practices. According to BirdLife International (1996), this poses a specific problem with respect to the headage payments under the CAP livestock regime, which competes with measures designed to achieve environmentally acceptable livestock densities. Brouwer and van Berkum (1996) cite the sheep premium, which leads to overstocking and overgrazing in parts of Ireland. The incentives offered under the agri-environmental measures are obviously not high enough to attract farmers into the environmental scheme. Competing support measures thus seriously limit the effectiveness of agri-environmental measures. Another example which restricts the effectiveness of environmental schemes is cited by Lampkin (1996). He claims that farmers wishing to convert to organic farming methods would experience a substantial loss of subsidies if they did so.

Some instruments, in particular the basic premiums granted in some programmes (France, Austria, Bavaria and so on), are no less controversial. Critics claim that a substantial share of the available funds is spent without attaching demanding environmental conditions to the premiums. There is probably some truth in this. Defenders of the basic premiums, however, claim that they do raise a general awareness among farmers of the need to protect the environment. The Commission hopes to obtain a deeper insight into the environmental effects of the various measures from a number of studies recently commissioned on the initiative of the Commission. The first results are expected in 1997.

Problems are posed also by the poor uptake of the Regulation 2078 programme in some MS. Existing experience shows that the acceptance of the Regulation 2078 measures is generally high in the northern MS, whereas it is rather low in southern MS. Three explanations for this phenomenon seem possible:

- Premiums are not attractive enough to encourage farmers to join a programme. There are two aspects here to take into consideration: firstly, it is important to know that average yields were taken as a basis for the calculation of the maximum cofinancing rates in Regulation 2078/92. Because of this average, the premium may not be attractive enough for farmers in favourable locations. As a consequence, programmes may not succeed in attracting some of the particularly polluting intensive agricultural holdings. Secondly, due to the need to cofinance, poorer countries may not have enough national resources to provide their share, so premium levels are lower than they could be.
- Some farmers hesitate to commit themselves for 5 years, the minimum period for participation in the programmes.
- Information to the farmers about the programmes is insufficient.

Recalling the objectives of the agri-environmental measures, such as to encourage farmers to reduce the use of fertilizers and pesticides, to use organic farming methods, to reduce livestock densities and to preserve water resources, natural habitats and the landscape, one might conclude from this analysis that, although there has been some progress, much more needs to be done in order to make CAP more sustainable. In this context, it should not be forgotten that the agri-environmental measures, with a budget of about 3% (1.4 billion ECU in 1996) of the overall agricultural budget (41.2 billion ECU in 1996), are hardly more than a fig-leaf, which is likely to be overshadowed by the environmental impact of the remaining 97% of the CAP budget.

PERSPECTIVES[2]

To conclude, some possible perspectives of the future development of the CAP will be highlighted.

- For the near future, quite a clear picture is already visible: firstly, budgetary cuts of about 2.5 billion ECU are to be expected. These cuts will mainly be at the expense of agricultural and regional policy. However, the agri-environment measures will probably not be seriously affected.
- Beyond that, a number of reforms of individual market regimes are to be expected, in particular in the fruit and vegetable sector, as well as in the

wine and olive-oil sectors. The Commission proposals presented to Council contain specific requirements for environmental protection. To what extent these will be approved by the MS is uncertain.

- The Commission recently produced a proposal for the amendment of Regulation 2092/91 on organic farming (COM(96)366 final). This amendment contains specific regulations and requirements for environmentally friendly livestock production. With this, a substantial gap in the previous regulation will be closed.
- In close association with the efforts to improve marketing of high-quality, ecological products, the Commission is thinking about the merits of introducing an eco-label for foodstuffs. With a view to solving the problem of an increasing and confusing number of so-called 'eco-labels' in the food sector, the Commission is considering the development of uniform minimum standards for the allocation of such eco-labels in the food area. As a first step, the Commission has issued a call for tender for studies on an evaluation of existing labels.
- Concerning the longer-term perspectives of the CAP, the considerations of the Commission are still in a very early stage.

The debate on the perspectives of CAP must certainly be seen first within the global, worldwide framework, with the 1992 Rio Conference and the Uruguay Round as the cornerstones. In particular, the next WTO round will be decisive for CAP. Under the existing agreement on agriculture, the WTO members undertook commitments to reduce internal support, tariffs and export subsidies within the implementation period of 6 years. Non-product-specific payments ('green-box' measures) and direct payments under production-limiting programmes introduced under the 1992 CAP reform ('blue-box' measures) are exempted from the reduction commitments. As Agricultural Commissioner Fischler pointed out in his statement before the joint meeting of the European Parliamentary Committee on Budgets and the Committee on the Environment on 16 July 1996, the EU is complying with its commitments, so no problem should arise within WTO at the moment. However, as negotiations for continuing the reform process in agriculture are to be initiated 1 year before the end of the implementation period and, as discussions in international forums, such as the Organization for Economic Cooperation and Development (OECD) or the Commission on Sustainable Development, already show, it can be expected that the main exporting countries of agricultural products ('Cairns Group') will put CAP under scrutiny. The Commission takes the message received from these discussions very seriously. They indicate that specific payments which are not sufficiently decoupled and for which there is no proof that the main purpose of the measure is a serious pursuit of social, environmental or rural development objectives may come under increased pressure. The Commission will further increase its efforts to evaluate and monitor programmes to ensure that the measures comply with the requirements of WTO obligations (Fischler, 1996).

Pressure on the agricultural policy also originates from increasing budgetary pressure. On the basis of increasing global demand for agricultural products, in particular cereals, high prices can be and are currently achieved. Costly compensatory payments can no longer be justified in the light of high world-market prices. Falling prices may require an increase in export subsidies. Finally, it is foreseeable that the envisaged EU enlargement to Eastern European countries will finally render the current system financially prohibitive.

With the background of these prospects, the Commission in December 1995 produced a strategy paper (*Agricultural Strategy Paper*), which discusses three possible scenarios: 'status quo', 'radical reform' (complete abolition of the price support, uncoupling of the compensations of production numbers, full remuneration of environment performances) and 'consolidation of the 1992 reform'. The paper rejects the 'status quo' and 'radical reform' in favour of an expansion and consolidation of the 1992 reform approach. The objectives outlined in this scenario are threefold:

1. Increase EU competitiveness on the world market.
2. Development of an integrated concept for rural areas.
3. Simplification of regulations.

Implementation of objective 1 suggests a further reduction of the price-support measures, in particular in the cereal area. This may be coupled – where necessary – to compensation. On the basis of the assumption that direct income transfers for social reasons ('blue-box' measures) will increasingly come under pressure, it is to be expected that the compensation will be linked more closely to additional environmental requirements in order to be compatible with the so-called 'green-box' measures under the WTO agreement. Green-box measures (aid decoupled from production, such as for environmental reasons) enjoy a much higher acceptance under WTO. Measures of a Regulation 2078 type will thus presumably gain in importance. At the same time, it is to be expected that the cost-efficiency and environmental effects of the agri-environment measures will become more important, as they will be critically scrutinized by our partners in WTO.

Implementation of objective 2 may result in an increased importance of environmental protection and sustainability in regional policy. The LEADER projects might have a model character here. To strengthen the multifunctional character of the rural area is a primary goal of Agricultural Commissioner Fischler. Such an integrated concept would imply that the environmental services rendered by the rural area, such as preservation and care of the cultural landscape, creation of attractive recovery possibilities and so on, would be remunerated more appropriately in the future. An important component of this concept will be the promotion of the manufacture and marketing of regional quality products.

Objective 3 implies that the regulation density and depth at Community level may be reduced in favour of a larger flexibility on the regional level.

European Union legislation could be limited more frequently than at present to the drawing up of framework directives and development of common objectives, as well as minimum requirements for measures to comply with the internal market. However, a certain measure of control by the Community will be necessary in order to prevent distortions of competition and to monitor the proper implementation of the programmes. The Austrian Memorandum for Mountain Agriculture ('Bergbauernmemorandum') discussed recently by the Council could be a model for granting greater flexibility to the MS in regional and structural policy.

On the background of the bovine spongiform encephalopathy (BSE) crisis, which will most probably incur costs equalling the overall budget available to the agri-environmental measures of 1992–1997, the perspectives of a sustainable agricultural policy will ultimately depend on the question to what extent the detailed reform proposals will be sufficient to solve the environmental problems caused by agriculture and deliver the environmental benefits expected by the non-farm population. In order to attain a positive balance sheet for the environment and for farmers as well, huge efforts are needed, including a close dialogue between administrations, farmers, environmentalists and scientists.

NOTES

1 The views expressed in this contribution are those of the author and do not necessarily reflect the views or policies of the European Commission.

2 The author wishes to stress that these views are strictly personal and do not necessarily represent the official view of the Commission.

Appropriate Frameworks for Studying the Relationship between Agriculture and the Environment: a Question of Balance

2

Julie Whittaker

INTRODUCTION

The biological basis of the production system is the defining characteristic of the agricultural sector but the nature of technological development has been to loosen the perceived biological constraints on output. External sources of fertility and energy have been introduced, and chemicals employed to control the 'vagaries of nature'. Such changes have disturbed established agroecosystems but, because of reduced dependence on the local environment as a source of inputs, the focus of concern has been on the loss of wildlife and the health risks from pollution. None the less, ecologists warn that, as production systems have become more specialized and dependent on inorganic inputs, long-term agricultural productivity is jeopardized. In some quarters their counsel has been taken, with some ecologically sensitive agricultural systems being developed, but not yet on any significant scale. The purpose of this chapter is to explore the implications for the agricultural sector of adopting an ecological perspective and, in so doing, to make a comparison between the framework of neoclassical economics and that of ecological economics. For clarity, the two frameworks are presented as distinct entities, but it is recognized that much contemporary economic analysis lies between the two extremes.

DISTINCTIONS BETWEEN NEOCLASSICAL AND ECOLOGICAL ECONOMIC FRAMEWORKS

A basic premise of neoclassical economics is that human happiness is derived from the consumption of goods, while processes of production are merely a means to this end and not sources of utility in themselves. The purpose of economic activity is to maximize output, given limited resources, and the purpose of economic analysis is to provide guidance towards this end. Decentralized choice, expressed through the market, is taken to be the most effective means of maximizing social benefit within an economy, except in cases of market failure. Since property rights are not clearly defined for many environmental resources, there are frequent instances of market failure pertaining to their use and government intervention is usually advocated. This is so for the agricultural sector which is considered to generate both positive and negative environmental externalities. Examples of the positive are wildlife habitats and pleasant landscapes, while polluting emissions are negative. Policy intervention, in the form of delineating property rights, assigning taxes, subsidies or tradeable permits, is deemed the appropriate means to correct the imbalance in the output of agricultural goods and environmental quality.

Ecological economics, while recognizing the needs and aspirations of the human race, places greater emphasis on understanding the interactions between economic and ecological systems. Whilst neoclassical economics has tended to view environmental issues as incidental cases of market failure, ecological economics perceives the economy as an open subsystem of the ecosystem and consequently intertwined with it. Enquiry into the ecological foundations of economic activity has grown in response to increasing recognition of the pervasiveness of environmental problems and concerns about sustainability, and in 1987 the International Society of Ecological Economics was founded, with the central objective of fostering sustainable systems (Costanza *et al.*, 1991).

Acknowledgement that the economy is a subsystem of the ecosystem requires an understanding of the latter in economic analysis. Ecosystems are complex structures of living organisms and biogeochemical and physical processes, which evolve over time. It is explicitly recognized that, because of complexity, the nature of the interactions between factors is not fully understood, but it is known that disturbances to ecosystems can lead to effects that are removed from cause by both time and space. As a result, it can be difficult to predict the consequences of actions and frequently the effects are not known until irreversible damage has occurred. For example, organophosphates have been found to cause health problems years after their use had been advocated. Similarly, the buffering capacity of the soil is already exhausted by the time visible effects of acid rain on flora and fauna have been noticed. These are cases where chemical inputs have dislodged an ecosystem. Problems can also arise when human induced changes to land use lead to increased vulnerability to natural forces, such as disease and extreme weather, which can dislodge the system from its self-organizing state. If catastrophic change is to be avoided, ecologists advise that ecosystem resilience should be maintained. Ecosystem resilience is the degree of disturbance that can be absorbed by an ecosystem before it loses self-organization and flips from one equilibrium to another. Unlike the parameters of many neoclassical models, ecological change is not always gradual and continuous and does not always permit a return to a previous equilibrium. The susceptibility to flipping is related to the displacement of locally evolved biota and a reduction in biodiversity, although there is not a strict correlation between increasing resilience and greater biodiversity (Holling, 1986; Holling *et al.*, 1995). Change in the semiarid grasslands in east and south Africa provides an example of an ecosystem flip (Perrings and Walker, 1995). The introduction of cattle ranching has changed the pattern of grazing from intense periods separated by intervals that permitted grass recovery to a moderate but persistent level of grazing. This new management system gives an advantage to drought-sensitive grasses at the expense of drought-resistant varieties. The result is that, now, droughts which were previously sustainable can cause the system to flip, with the land becoming dominated by woody shrubs. Unlike non-renewable resources, for which rising scarcity is indicated by a gradual rise in price, allowing time for technological and institutional adaptation, the sudden change in ecological processes can result in substantial economic disruption. Ecological economics recognizes that gradual ecological change is a normal part of evolution but emphasizes the need to avoid catastrophic change in order for development to be sustainable (Arrow *et al.*, 1995).[1]

Temporal considerations, particularly longevity, are central to any analysis concerned with sustainability. Such concerns are not easily embraced within the neoclassical economic framework, which is more suited to static analysis (Batie, 1989). Its mechanistic model-building approach allows only a limited number of variables, yet, once time considerations are introduced, less constancy can be realistically assumed. Sometimes conjectures are made on the direction of preferences and technological change, typically on the basis of extrapolation from trends. As a result, predictions for the long and medium term are rarely accurate because they deny the complexity which exists. Ecological economics gives greater acknowledgement to the fact that economic systems, like ecological systems, are complex. Not only can there be prolonged ecological feedback effects that are unpredictable, but preferences and technology evolve in a manner which is not necessarily an extension of past trends. Instead, it stresses that preferences and technology are in part endogenous to the system, although, as with genetics, the possibility for novelty exists. Norgaard (1992a) argues that development should be considered as a process of coevolution between knowledge, values, organization, technology and the environment:

> deliberate innovations, chance discoveries, and random changes occur in each subsystem which affects the distribution and qualities of components in each of the other subsystems. Whether new components prove fit depends on the characteristic of each of the subsystems at the time. With each subsystem putting selective

> pressure on each of the others, they co-evolve in a manner whereby each reflects the other.
>
> (Norgaard, 1988)

Probably economists from many persuasions would have little quarrel with this statement; the differences arise with regard to the interpretation of the implications for research.

Environmental economics acknowledges some of the specific issues raised by environmental problems but is restricted in the extent that it can address them, in so far as it maintains allegiance to the principles of the neoclassical framework. These principles are that theories should be based on the assumption of rational agents and they should be open to testing by statistical inference.[2] This requires reductionism. Certain quantifiable components of the economic system are identified as being important determining variables, while other components are assumed to be constant. For example, economic factors that cause changes in demand, such as prices and incomes, are allowed to vary, but other variables that are relevant to a long-term perspective, such as preferences and technology, are assumed to be exogenous and, in many cases, undeviating, because they are unquantifiable and thus not amenable to statistical testing. The formation of preferences and, to a lesser extent, technologies is therefore ignored by the neoclassical approach, because of the difficulty in developing theories that are statistically testable. Yet the long-term development of preferences and technologies is a crucial determinant of ecological and economic sustainability and some understanding needs to be developed, if by scholarship rather than statistical analysis. The criticism that the neoclassical approach fails to address the fact that economies are open dynamic systems has also been made in a more general context (Hodgson, 1988; Lawson, 1995; Dow, 1997).

Although neoclassical economics has brought valuable insights to the analysis of environmental problems, its methodological constraints limit comprehensive analysis. Meanwhile, ecological economics is developing within a more liberal methodological framework; indeed, it professes to be methodologically open and to encourage the integration of the study of ecology and economics, in the belief that this will result in more effective policies (Costanza *et al.*, 1991).

The long-term perspective of ecological economics and its appreciation of evolutionary change have led some commentators to suggest that ecological economics is Darwinian in character, in contrast to neoclassical economics, which, in tending to focus on static analysis, has more in common with Newtonian method (Batie, 1992). From a theoretical perspective, this may have some relevance, but it is pertinent to note that Darwinian thought has been used by exponents of the neoclassical approach to justify market outcomes. For example, the increasing concentration of firms within a market has been absolved on the grounds that it exemplifies the survival of the fittest. Obviously a system driven by atomistic individual agents will lead to some evolutionary outcome. The doubts raised by ecological economics is that this will not be a sustainable one.

THE IMPLICATIONS OF THE ECOLOGICAL ECONOMICS PERSPECTIVE

The ecological economic framework stresses the ubiquity of the environmental effects of economic activity and the importance of maintaining ecosystem resilience to ensure sustainable systems. It might be said that the ecological approach moves towards the deontological position that in all decisions the ecosystem should be respected. An implication of this is a wider appreciation of the laws of thermodynamics, which advocate less dependence on non-renewables and more dependence on renewables (Georgescu-Roegen, 1979). This is predominantly because the use of non-renewables causes the dispersal of matter, which leads to pollution and ecological disturbance. A greater reliance on renewables would return economies to more cyclical patterns, whereby waste matter decays and provides nutrients for the growth of further resources.

Ecological economics notes that economic activity determined by individuals making choices independently can lead to ecological instability (Berrens and Polansky, 1995; Common, 1995) and thus have an impact on welfare. Moreover, government intervention cannot be guaranteed to provide a sufficient guard, because of the pervasiveness of the potential problems, the high informational requirements associated with a top-down approach, and the likely reluctance of governments to advocate adherence to ecological principles when cause and effect cannot be identified precisely. A recognition that environmental problems are pervasive, at a time when the globalization of economies is constraining public expenditure and placing industries in Western countries under increasing competitive pressure, signifies that policy design for sustainability needs to be not only ecologically tolerable but also economically viable. Thus, it will need solutions beyond the conventional policy options. What is required is not a few minor adjustments to the general direction of progress, but an expressed intent to indicatively plan the orientation of economic activity. Consequently, it is suggested that there is a need to involve all parties in the design and organization of socio-economic systems in order that commitment to maintaining ecosystem resilience is maximized. A collaborative approach could also have the advantage of resolving some conflicts and thereby reducing some opportunity costs (Costanza *et al.*, 1996). This approach is at odds with the principles of the neoclassical framework, which sanctions the actions of self-motivated individuals driven by their own preferences and extends this principle to environmental issues, where individuals' willingness to pay is sought. Deference is given to the concept of a consumer-propelled economy and the supposed freedom this confers on the individual. Yet this position ignores the fact that choice is limited by the structure of the economy and that consumers acting as individuals are powerless to change the economic structure. If the political process adheres to these economic precepts (and this has increasingly been the case over the last two decades), the economy moves on without steerage.

The lack of a social context within the neoclassical framework is berated by Heilbroner and Milberg (1995). They express the need for a 'vision' – a set of widely shared political and social preconceptions – within which economic analysis can operate, thereby replicating the classical economic approach. Thus, they echo Schumpeter (1954, p. 41), who wrote that 'analytic effort is of necessity preceded by a preanalytic cognitive act that supplies the raw material. In this book, the preanalytic cognitive act will be called Vision.' Costanza *et al.* (1996) assert the need for a practical shared vision both of the way the world works and of a sustainable society, as a vital element in achieving a sustainable system. It might be argued that the development of a vision is the role of the political system, but an endorsement by economists that this is required, rather than explicitly or implicitly underwriting unfettered consumer sovereignty, would in itself be useful. However, it should be borne in mind that political systems (in Western countries) have typically induced only incremental change, usually for short-term rewards. The development of a vision for a sustainable system requires a holistic and long-term approach and may require new social forums.

FOSTERING SUSTAINABLE AGRICULTURE

There is a case for developing a vision for the agricultural sector instead of allowing it to evolve in its current *ad hoc* way. Technological development is largely unguided, and usually only when the detrimental effects emerge is a corrective policy instigated. There could be efficiency gains from specifying the type of agricultural system that is required and directing technology towards that end, rather than following a strategy that amounts to bolting on repairs as the weaknesses emerge. Moreover, a design strategy provides greater choice than the current incremental policy approach, which, in the case of environmental measures, often seeks justification through willingness-to-pay studies. But with this technique only marginal changes within the status quo can be expressed. In addition, the neoclassical framework assumes separability between outputs, with individuals forming their marginal values for agricultural products and environmental quality independently of each other, although in reality this is not likely to always be so. Consider the case of traditional land-management schemes. When valuation studies are conducted to ascertain the social value of a land-management scheme, the choice is for the land to be managed according to traditional methods or to be developed in whatever direction modern technology dictates; there is no opportunity to explore other options. However, a recognition of ecological limits and an opportunity to express these concerns could lead to a questioning of the current tendency in policy (at least within the UK) of conserving particularly valued agroecosystems, while the majority of agricultural land is managed with little reference to ecological principles. A more open forum would permit a widening of the debate to address the required balance between food production, the preservation of existing habitats and the more extensive promotion of ecologically sensitive farming practices. It is possible that, if farming systems based on inorganic inputs were to become less prevalent, then total output might decline. A question which would require addressing is whether lower levels of domestic production are acceptable or whether means of intensifying production on some of the currently preserved areas in ways that do not compromise ecosystem resilience are preferable. All aspects need to be addressed together, rather than incrementally.

The need for these choices to be made has arisen because agricultural technology has developed in such a way that it has had adverse side-effects. Pretty (1995) argues that technological change has been driven by an agricultural science which, in common with mainstream economics, has been dominated by a reductionist approach. Environmental problems have emanated because agricultural scientists have oversimplified intricate systems. Purchased inorganic inputs have been favoured because they are more 'controllable' than organic inputs. As a consequence, specialization has been encouraged and production systems have become more linear and less cyclical, returning less waste matter to the production cycle. Longworth (1992), in his presidential address to the International Association of Agricultural Economists, suggested that much agricultural research over the last 40 years has been misdirected. This has resulted because the reward system for agricultural researchers (and the same may be said for economists (Rosenberg, 1992)) has encouraged disciplinary rigour while complexity has been disregarded. He suggests that 'the long term multidisciplinary non-scientific features of the problems involved [with sustainable development] "frighten" young ambitious and capable scientists'. But it is not only the institutional structure of research establishments that have conditioned technological progress but also the market structure. The existence of the agricultural-input industries is dependent on generating innovations that can be embedded within a purchasable input.

Agricultural practices that are more compatible with the ecosystems within which they function will require natural assets to be utilized rather than suppressed. This requires the development of a diversity of agricultural systems to suit varying local conditions. Norgaard (1992b) questions whether government agricultural departments are able to facilitate such diversity and counsels the need for decentralization. Bawden and Ison (1992) and Pretty (1995) argue that farmers need to be involved in the development of agricultural practices if these systems are to become more appropriate. Local involvement is central in devising new methods, not only because the local population has an in-depth knowledge of local environmental conditions but also because new practices will be adopted only if they are acceptable to the farmers, who are themselves part of the agroecosystem. Bawden and Ison (1992) and Pretty (1995) consequently suggest that agricultural development should be considered as a learning process that involves all those who are part of the system. This requires a change in the role of agricultural research establishments, from purely disseminating 'expert' knowledge to become learning organizations promoting experimentation and connecting with local farmer groups. This concept of participatory development is more familiar in tropical and subtropical regions than in temperate climes, perhaps because the appropriate nature and process of development are an issue in the former regions, whereas in Western countries, where higher levels of output and income have been reached, the nature of further growth is commonly assumed to require little direction.

In finding economically viable paths to ecologically sustainable systems, encouraging individuals and organizations to seek mutually beneficial arrangements that operate within ecological constraints is an appealing approach. This strategy of encouraging win–win outcomes has its successes (Greyson, 1995), but Wolf and Allen (1995) counsel that the inevitability of some trade-offs and the consequent need for social choices to be made have been insufficiently addressed by proponents of sustainable development. It is possible that this precedent was set by the Brundtland report (World Commission on Environment and Development, 1987), which shrewdly gained popular support for the concept of sustainability by suggesting that economic growth was acceptable, indeed a necessity, in a shift to a more environmentally safe system (Common, 1995). Ecologists, in association with economists, recognize that trade-offs do exist, but there are differences in their means of analysing and handling them.

Neoclassical economists confront trade-offs by weighing up the costs and benefits of change. Within the ecological approach, hierarchy theory, which illustrates the trade-offs that exist within the natural world, has been extended to consider human interactions (Allen and Starr, 1982; Conway, 1987). Ecologists view ecosystems as a nested hierarchy of systems; for example, individual plant and biogeochemical processes occur at one level, while above are animal and abiotic processes, and above them are geomorphological and global biogeochemical processes. Within these three broad systems, there are further nested scales. For one scale of processes to adjust and survive, it can be necessary for a lower, smaller system to die. Evolution depends on the limited life of the component parts, so that new options can be selected. 'Sustaining life requires death', state Costanza and Patten (1995). Similarly, it is not possible to sustain everything within the economic subsystem; some trade-offs will be inevitable. Wolf and Allen (1995) argue that this needs to be recognized and addressed if sustainable agricultural practices are to gain widespread acceptance. They elucidate that, while the concept of sustainability is scale-independent, that is to say, it can be applied across a wide range of spatial and temporal scales, for analytical and policy considerations it is important to set bounds to the problem by making it scale-dependent. Once the scaling is clear, the links between scales can be made, and the possibility of having to sacrifice sustainability at one level in order to achieve sustainability at a higher level can be explicitly considered. Wolf and Allen (1995) suggest that improved institutional arrangements are required for addressing the inherent trade-offs in agricultural systems. They hold that complex, multilevel systems cannot be organized with simple optimization techniques, because they depend upon too high a level of aggregation and thus hide important information. Instead, they advocate that the selection of trade-offs requires a holistic management approach. This requires identifying, assimilating and acting on information in a manner that allows the integration of multiple considerations across all levels in the system.

ARE THE TWO FRAMEWORKS COMPETITIVE OR COMPLEMENTARY?

The conventional approach to studying the environmental impact of the agricultural sector views the issue as one of balancing the competing outputs of agricultural and environmental goods, while the focus of the ecological approach is to achieve balance within the ecosystem so that it is sufficiently resilient to withstand shocks and ensure long-term productivity. Are these two frameworks mutually exclusive or do they complement each other?

From a methodological position, the two frameworks are very different. The neoclassical approach has a restrictive methodology, while the ecological approach adopts methodological pluralism. The latter emerges from a recognition of the intricacies in ecological and economic systems and that there are multiple realities which are socially constructed. Therefore, there are limits to generalizations, but this does not deny a role for theories and statistical inference. The acceptance of complexity and the associated uncertainty (for which there are limits to reducibility) leads to an attraction to policies that emphasize a cautionary approach. Although a consequence of this is to introduce more ecological constraints into an economic system, the evolutionary perspective and emphasis on the social context of preference and technology formation suggest that the constraints may not be as binding as conventionally thought.

In practical terms, in finding ways of understanding the world and in encouraging sustainable systems both frameworks have their place. The issue is to define each framework's area of usefulness. Bawden and Ison (1992) suggest that a holistic framework, which the ecological approach provides, is appropriate as a starting-point for looking at a system. More reductionist methodologies are applicable for addressing the specific problems that are identified by the holistic analysis. With respect to agriculture, they stress the importance of involving the participants of the agricultural system (these can include not only farmers but also consumers, countryside users, researchers, policy-makers) in the initial stage, because all learn from the experience and gain a better appreciation of how the system works and this often leads to a change in values. The reductionist models that the neoclassical framework offers have a role, once the parameters of the problem are definable and the goals are clear. However, since reductionist models are limited by definition in the number of objectives they can analyse, they can be but one input into the decision-making process. Issues of longevity and resilience, which are central to sustainability concerns, cannot be captured easily within such models, although maximizing subject to ecological constraints may be appropriate in some instances. The ecological perspective provides a specific challenge to policy-makers in stressing the importance of ecological resilience. What determines resilience is still an area of debate among ecologists (Holling *et al.*, 1995), but it is relatively clear that diversity between regions and diversity of biota within regions have a bearing. Uniformity has typified policies in the past, in part because they are so much easier to promote from the top down, but such uniformity has been a source of environmental problems. A means of encouraging diversity is by decentralizing decision-making and involving local populations in the development of appropriate agricultural systems for their locality. This is consistent with the view that there are multiple realities which are socially constructed. Participants hold the expert knowledge of the local environment, thus permitting the natural assets to be more fully utilized, and the variety of experience involved will encourage more varied and original solutions. This does not negate the need for researchers to foster such developments and proffer advice, but the emphasis is on encouragement and cultivating a learning process rather than suggesting a completed design. Nor does it negate the need, in certain circumstances, for taxes or subsidies when otherwise locally defined goals differ from those of the wider society. Economists have a role in exploring the institutional arrangements that will stimulate sustainable agricultural systems. These could range from devising incentives to induce farmers to give time to a participatory approach to considering the ways in which tenure arrangements affect the planning horizon.

CONCLUDING COMMENTS

The reductionist methodology of neoclassical economics is too constraining to permit a full analysis of the factors determining agriculture's relationship with the environment. As a consequence, it can limit choice and lead to inefficiencies by not appreciating the wider picture. Moreover, it gives little consideration to ecological resilience, which is essential for the long-term sustainability of agriculture. In contrast, the ecological economic perspective provides a broader framework, by offering a systems approach, and it explicitly acknowledges the significance of ecological resilience. However, the holistic perspective of ecological economics requires contributions from many sources, and neoclassical economics, in association with other branches of economics, can offer valuable insights. For example, it can clarify the nature of the choices that society faces by evaluating some of the opportunity costs and it can elucidate on the insurance value of ecological resilience. Once choices have been made, again the economist has a role in designing market mechanisms to achieve stated ends and in suggesting appropriate institutional arrangements for encouraging sustainable agricultural development. Both frameworks have their purpose, each is dependent on the other; the determination of their respective roles is a question of balance.

NOTES

1 It is true that the concept of safe minimum standards, devised by Ciriacy-Wantrup (1952), which advocates taking caution when outcomes are irreversible and potentially significant, is accepted by many neoclassical economists. Differences in opinion lie in what actions are likely to lead to adverse significant outcomes.

2 It is difficult to find a term that adequately describes the prevailing methodology in mainstream economics, in part because there is no open debate on this (Dow, 1997). Some commentators use the term 'positivism', even when noting Popper's contributions; others use 'modernism', but there is no consensus. For an excellent synopsis of methodological developments in economics, see Randall (1993).

The Po Delta Park: One River, Two Policies

3

Giorgio Osti

INTRODUCTION

In Italy, the delta of the River Po, considered in historical and not strictly hydrological terms, extends over an area of about 145,000 ha, divided almost equally between the Veneto region and the Emilia-Romagna region (Tomasin, 1990). It is the most important wetland area of the country, reported as a zone deserving special protection by the International Convention of Ramsar in 1971.

For 30 years, there has been talk of creating a natural park in this area, and in the last 10 years pressure to establish the protected area has increased. The outcome has been different in the two administrative regions: one has started to create a regional park, while the other is still engaged in laborious bargaining, which, for the moment, has not given rise to any improvement project. The different outcomes of the park issue in the two regions provide an opportunity to assess the importance of political and social explanatory factors. The two sides of the delta area are, in fact, quite similar in their geographical and historical features, but they differ in terms of their political relations to administrative centres.

From this perspective, the actions of farmers and their associations with regard to the park project will be evaluated. The attitudes of the 'agricultural world' to environmental policies are certainly affected by very large-scale factors, such as European Union (EU) action and international market trends. However, this does not exclude local factors, which also influence events. In this case, one may hypothesize that the relationships between agricultural associations and local authorities have affected the outcomes of environmental policy in the Po delta.

The establishment of a natural park is a good area in which to examine new trends in the agricultural world's attitudes towards society, in a period in which its political strength is dwindling. The Po Delta Park, more than other environmental policies, is global in nature; it is a constituent policy (Lowi, 1972) and requires the large and detailed restructuring of the social roles of numerous actors, including farmers.

This chapter is organized as follows: the first part outlines three theoretical points within which to frame the society–farmer relationship; the second analyses the history of the Po Delta Park, using the theoretical frame; and the third part draws some tentative general conclusions.

THE RELATIONSHIP BETWEEN THE AGRICULTURAL WORLD AND SOCIETY

The relationship between the agricultural world and society has changed profoundly in recent years. The balance between agriculture and society, established in the 1950s, when the European Community was born, has broken down. That pact envisaged agriculture as playing an important role in the wealth of countries, by guaranteeing a certain food supply and by giving social stability to local communities affected by rapid economic modernization (Brunori, 1994). In that scenario, the agricultural function was eminently social: it was responsible for the security of the whole of society.

Although this role has not disappeared, it has been greatly curtailed by recent events. The enormous increase of productivity and market exchanges has eased the problem of food self-sufficiency. The agricultural population has diminished in relation to the number of rural inhabitants; consequently, the farmers' role as social stabilizer has become less crucial. It has also been demonstrated that farmers bear some responsibility for the environmental crisis, and they must find some remedy if they do not want to harm the population as a whole. In conclusion, these three processes have provoked a crisis in the overlapping of agricultural world interests with social interests.

This identification of interests used to engender specific ways of governing the agricultural world. Many scholars agree in defining the relationship between farmers' associations and public authorities as 'corporatist' (Lehmbruch, 1984; Cox *et al.*, 1986; Gueslin and Hervieu, 1992). The corporatist relationship is an exchange between the public administration and the agricultural interest group which is beneficial to both of them. The public authorities, generally, ask for consensus and technical–managerial know-how; the farmers' unions ask for public funds and the power to influence decision-making in the agricultural sector. A public–private decision-making pool is thus created, the features of which are a few members, stability and closure to external pressures (Lange and Regini, 1987). This kind of relationship arises under particular conditions, many of which depend on the internal assets of farmers' unions – unity and homogeneity, for example.[1] But other conditions concern the social climate: indeed, corporatist relations become stable and efficacious when there is broad consensus on the importance and specificity of the agricultural sector. In other words, the corporatist relationship succeeds when the entire society recognizes that the sector has an important function to play for everyone concerned. It is then that authorization to govern the sector is given to a restricted and closed group. This is possible because there is broad consensus; the corporatist pact can thus manage the technical aspects.

When the social function of agriculture is not clearly identified or when there is a suspicion that the sector is damaging to the public interest, the corporatist model enters crisis (Farago, 1985). Technical management loses its character of neutrality and becomes a political issue. The decision-making group changes from a policy community to a policy network (Frows and Tatenhove, 1993). The hierarchies are flattened; new actors enter the political arena or try to do so; issues shift from the technical dimension to the political one. Problems of ultimate aims and the distribution of benefits are raised. The decisional arena becomes crowded and chaotic.

The events linked to the environmental question lend themselves well to interpretation on the basis of this model. Agriculture, like other productive sectors, has been charged with having a heavy environmental impact. The corporatist balance has been stressed. New actors – public and private – have sought to enter the agriculture decision-making arena.[2] The negative reactions of the agricultural world to the pressures on it to comply with environmental standards can be understood not only as fear of income loss but also in political terms: as a crisis in the traditional way of framing the relationship between farmers' unions and public bodies, between agriculture and society.[3]

A second point of relevance to the understanding of the difficult agriculture–society relationship is the following. Since the end of the Second World War, the Italian agricultural world has pursued the principal objective of close integration with the market. This was seen as a means of helping peasants to escape social and economic marginalization, giving them dignity with respect to other professions and enabling them to earn an adequate income. This effort, after vertical integration, absorbed most of the energies of the unions and cooperatives and it also helped farms to become more competitive. However, the endeavour also brought social costs. Benvenuti (1975) was the first to signal the loss of farmer entrepreneurship caused by the progressive absorption of many tasks by external agencies. The farmer has thus become more a part of the system than an actor. The loss of entrepreneurship can be viewed as a form of alienation from work. In social terms, the highest cost has probably been the abandonment of local communities. For a long time, the peasant and rural communities were identical. But the cleavage between the two was not only due to the decline in the number of farmers; market integration was also responsible. The *filière* (chain) integration imposed market centres localized very far from the local community as the main territorial reference. The farmers' unions encouraged this integration because it helped the peasant to become a rational producer. In so doing, however, they forgot the territorial interests which had pivoted on agriculture for centuries. Conflicts in the villages between organized farmers and the rest of the population arose because of the separation from local life.

Agriculture is not always organized in this way. Van der Ploeg and Saccomandi (1995, p. 14) describe and name two production styles, that of huge (or intensive) farmers and that of economical farmers. The first are closely dependent on an external technological model, strong market integration and high transaction and transformation costs (see also van der Ploeg, 1995). In contrast, the second kind of farmer embodies more technical skills; he/she is more closely tied to local production factors, particularly the land, and bears higher management costs. The huge farmer has lost most of his/her autonomy and most of his/her relationships with the surrounding milieu. Moreover, the close dependence of this style on external inputs means that he/she is probably less able to cope with increasing ecological constraints (van der Ploeg and Saccomandi, 1995, p. 15).

Basile and Cecchi (1995) draw a less dramatic distinction between models when they talk of two kinds of agriculture – the integrated (*omologata*) model and the non-integrated (*non-omologata*) model – which can coexist in the same area or even on the same farm. Integrated agriculture has modernized well because it has intensified the use of technologies and its ties with the market. Although non-integrated agriculture has the opposite features, it is not marginalized agriculture. It can survive alongside its modernized counterpart because it provides high-quality goods for farmers or customers particularly sensitive to environmental concerns. This kind of agriculture makes it easier to provide services – from the farmhouse to guest accommodation – which integrated agriculture is unable to provide because it is highly standardized and delocalized. Non-integrated agriculture is able to survive because of local opportunities to earn mixed income and because of the growing importance of environmental amenities for the tourist industry.

These distinctions are linked to the polarization between *filière* and district. A *filière* implies the integration of the production, processing and marketing phases of a good without reference to a particular territory, at least in the downstream phases. The agricultural district has close local integration among phases and environmental qualities distinctive of the goods produced and processed in the area (Di Iacovo, 1994). Real situations seldom belong clearly to one pole or the other; intermediate cases are most frequent. According to Di Iacovo (1994), the capacity to affect policies differs in the two production organizations: firms and institutions involved in the chain or *filière* are better able to bargain on specific issues and the bodies working in the final phases (market short side) usually gain more.

Those involved in the district are less ready to negotiate and less flexible, but they have the advantage of joining specific and general interests better (Di Iacovo, 1994, p.142). District-organized firms may be more suitable to deal with policies regulating public goods, such as environmental policies, because they are more able to match proposals and to offer stable delegations to the political decision-making arena.

In conclusion, the integration of the agricultural world with local society depends on the kind of product (for example, the cereals market is highly delocalized), the way in which the marketing of goods is organized (for example, whether it is done by cooperatives or by other bodies) and the productive vocation of the area (for example, a famous wine usually connotes the area's economy and culture). Real-life situations are highly variable and sometimes they are a mix of different styles and organizations. There is no doubt, however, concerning the importance of the territorial dimension: farms and the production of specific foods have different territorial references. When these references are outside local control, it is likely that the farmers' separation from the local community will be greater, too.

The third aspect to be considered in the relationship between the agricultural world and society is a pre-eminently cultural one. Since the end of the Second World War, the great mass of Italian land workers and farmers have lived through an epic, a huge undertaking: the fight for their own social redemption (*riscatto*). In this fight, they were flanked by other workers, often ideologically distant from them (for example, the communist workers in industry). The escape from poverty, the breaking of dependence on big tenants, victory over the environment: these were the ideals that inspired the worker and farmer generations that arose from the Second World War. And they were ideals that were typically modern. They stemmed from the impulse to free humanity from the chains of economic poverty, sickness and traditional authority. They were legitimate ideals because they were the shared heritage of the popular classes, which were enabled by universal suffrage to fight politically to achieve them. In conclusion, it can be said that a civil society arose around the mission for the social redemption of the popular classes. A common base was formed beyond distinctions of ideology and wealth (Alexander, 1995), a common base that was highly conducive to the functioning of political power: the representatives knew the people's ultimate aims very well.

The ideal of redemption has been largely fulfilled, although the social costs (for example, emigration or pollution) are not easily identified. Nevertheless, a certain amount of wealth, insurance guarantees, and technological modernization have been achieved by most farming families. Even if inequalities persist and the future seems rather uncertain, the grievous conditions have almost everywhere been superseded.

The mission for social redemption has undergone a more subtle change: the crisis of the moral and existential assumptions of modernity (Crespi, 1989). The main components of modernity – the individual, progress and rationality – are seemingly under strain in contemporary society. They are considered values that are outmoded, even harmful. For example, the environmental crisis is often explained in terms of modern humanity's attempt to achieve limitless development by means of rational tools. Moreover, individualism and rationalism restrict humanity's self-expression because they leave no room for sentiments and sociability. Every relationship is driven by instrumental motivations.

The agricultural world undoubtedly relied very closely on the entrepreneur (individual), who, by means of the systematic use of technology (rationality), sought to achieve constant improvement of his/her human and professional condition (progress). This image of modernity, on which the agricultural world spent so much energy, is now in crisis. The mission of social redemption, progressive liberation from poverty and nature, is no longer a unifying myth.

The three aspects discussed above – corporatism in the government of agriculture, the degree of delocalization in the farm–market exchanges and the myth of social redemption – help to delineate the agriculture–society relationship. According to these aspects, there is a general and prevalent trend of leaving society by farmers or towards greater farmer–society conflict. Part of this scenario is the project for the natural park. What reactions does it provoke in the agricultural world? Can it facilitate relationships within society or, conversely, does it exacerbate conflicts? And, if the latter hypothesis is more likely, why?

THE PO DELTA PARK: A DETONATOR OF AN AGRICULTURE–SOCIETY CRISIS

The natural park is an all-embracing project in two senses. It is global in a logical sense, because those concerned with environmental protection in a large area must necessarily be concerned with numerous other activities in that area. If the environment is like a system, such as a whole with interdependent parts, action taken in one part affects, directly or indirectly, all the other parts. It is an all-embracing project in a political sense, because the recent mainstream trend in environmental policies, sanctioned at the Rio Conference and at the 1992 Caracas Conference on parks, is to link them with territorial policies. In other words, environmental protection must be deliberately coupled with the social and economic development of the local population.

On these premisses, the park project assumes great significance. It can provoke profound changes in the balance of power, economic activities and cultural trends. It can alter numerous aspects of the farmer–society interaction (Billaud, 1986).

The following hypothesis can now be formulated: the proposal of a global park exacerbates agricultural world–society relations already made difficult by the issues outlined above. In other words, the park 'detonates' a crisis of identification between agricultural interests and collective interests. A new synthesis, a new overlapping, can be found only if farmers' interests and social interests are redefined. The Po Delta Park tests this hypothesis. Before analysing it in terms of the three-point frame, it is useful to describe the main events involved in its proposal.

In the 1980s, the Italians grew increasingly aware of environmental issues. Groups and associations increased their membership. A Green Party was born, which achieved relatively good results in general and local elections. Nevertheless, the ecological movement has been an urban phenomenon, as evidenced by the territorial distribution of votes and the results of referendums on environmental issues. The green vote is polarized between urban and rural, an unusual cleavage in Italian politics.

The Po Delta Park was long confined to the proposals of small specialist groups, environmentalists' programmes and to the land planning of the regional administrations. A great deal was said but nothing was done. Local people did nothing either for or against the park. The breakthrough came, first, with the institution by law of the Emilia-Romagna Regional Park (1988) and, second, with the protected-areas framework law (*legge-quadro*), approved by the Italian Parliament in 1991. The framework law established that one park was to be created in the delta areas of both regions.

In the 1980s the Emilia-Romagna region worked to convince the local people of the advantages of the park, achieving, after laborious negotiations, the institutive law of 1988. During the negotiations, the agricultural world was able to give voice to its interests and managed to have most farmland excluded from the park. The Veneto region, instead, was indifferent to a parks policy and to the Po Delta Park in particular. It was one of the northern regions with the least amount of specially protected land. This different attitude has probably been due to ideological reasons and political contingencies. The Christian Democrat Party, the leading party in the Veneto region, embraced a more *laissez-faire* ideology than did the Communist Party, dominant in Emilia-Romagna (Bagnasco and Trigilia, 1984). Moreover, the Venetian part of the delta was not in harmony with its region's political trends.[4]

Approval of the national framework law placed the two regions under pressure. Emilia-Romagna had to abandon its regional project and Veneto was forced to change its dilatory strategy. The framework law gave them the opportunity to undertake joint management of the protected area by means of an interregional park. However, the two regions never reached agreement on the form that such joint management should take. As a consequence, the Emilia-Romagna region decided to go it alone on the basis of the old regional law, while the Veneto region did nothing. The national government, on the other hand, was unable to create a national park, as envisaged by the framework law, unless the two regions reached agreement.

The different outcomes of the park affair on the two sides of the Po highlight agriculture–society relationships according to the three-point frame: corporatist relationships, the local integration of agriculture and the culture of social redemption.

The park doubtless threw the traditional corporatist order used in the government of agriculture into crisis. As a global policy of urban origin, it failed to take account of the interests of the agricultural world. In the first phase of the creation of Italian natural parks, the agricultural world was completely excluded. Only when the park model as the harmonious synthesis of humanity and nature came to predominate – such as when the purely regulatory version had been superseded – did some regions appoint farmers to the park management boards. However, the framework law again changed this practice by introducing territorial interests via local councils (such as municipalities, provinces and regions). In other words, as a result of the national framework law, farmers are not directly involved in park management boards. The environmentalists, instead, are included because they are considered to be representatives of a general interest.

As a consequence, the natural park, because of its origin and recent events, is far from the agricultural world. The farmers are regarded not as actors but as passive recipients of park decisions. At most, they are listened to during some phases of drawing up the park plan. The practice of sectoral corporatism, developed through agreements between farmers' unions and agricultural public boards, is not affected; it is simply excluded. Territorial management, if under the park's powers, has passed to public offices (environmental services) and to professionals (biologists, architects, naturalists) distant from the agricultural world. The farmers' unions cannot immediately and openly ask to be included in park management. Agricultural production is too exposed to the criticism that it pollutes the environment. Even if the farmers could show the opposite, they would find it difficult to present themselves as the defenders of general interests.

The corporatist agreement has been swept away by the park proposal. Rules and general interests have been redefined. In Emilia-Romagna, the farmers are consulted not by agricultural public boards but by local councils. It is not possible to re-establish the traditional exchange because the local council introduces more general than agricultural interests into negotiations. The farmers have managed to have their land placed outside the park borders. They will enter when they see their chance. Consequently, it is no longer possible to undertake the corporatist exchange of consensus for joint management. The farmers have 'won' because they will not be subject to the park regulations, but they have 'lost' because they can no longer play a powerful role in the management of an increasingly valued resource: the environment. In Veneto, they are not consulted in a systematic manner because of the ambivalence and uncertainty of politicians. In any case, the negotiations are difficult because of the farmers' opposition to any park proposal. They have even rejected the word 'park'.

The second frame point in the relationship between agriculture and society is the degree of local integration among productive functions. Agriculture in the Po delta is traditionally orientated to cereal and sugar-beet production. These are goods with low territorial specificity and they are indiscriminately stored in huge warehouses. The trade companies are very large in size and are usually located outside the delta. Market integration is very high. The territorial scale of processing and trading companies is very large. The example of sugar beet provides a clear illustration: there is a high level of sugar-beet production in the delta, while the large factories processing the beet do not have relationships with local firms and are owned by multinational companies (Brusco and Natali, 1992).

It is not difficult to show that this kind of market integration fails to induce the farmers to intensify local ties. The *filière* extends well beyond the local milieu and interests are defended to a very high degree. Integration with other local activities, such as tourism or artisanship, is low. Consequently, the farmers are involved more in a *filière* organization than in a district organization. The park, as a tool of territorial global policy, does not fit very well with this kind of vertical integration.

In recent years, the Po delta has seen the widespread development of horticulture. The small farms resulting from land reform (the distribution of land in the 1950s) are unable to survive with extensive production and have tried the route of specialization. In some cases, there is also a long-standing tradition of horticulture. This kind of production is more closely linked to the local economy. In horticulture, it is easier to exalt local vocations and often cooperatives are able to keep some *filière* parts local. In this situation, the park may be a good channel to promote local production, enhancing environmental qualities. On the Emilian side of the delta, there has been talk of using a park label for horticultural products. On the Venetian side, there is no mention of similar marketing operations: opposition to the park is so pervasive that no one, probably, would dare to make such a proposal. It is clear that the use of a park label depends greatly on local cooperatives, which at the moment do not possess the resources to exploit such a sophisticated mark as the park label.

The third frame point in the relationship between agriculture and society is the culture of social redemption. The culture of the Po delta has been strongly marked by two events: land reform and land reclamation. These two processes embody the idea of social redemption and redemption from nature. Farmers in the area see themselves as the agents of modernity. They and their ancestors fought against a hostile environment (flood, malaria), reclaiming the land from the water. Moreover, they fought politically to obtain better living conditions and land ownership. The land is not perceived as natural, not given by nature but wrested from nature. The land is kept dry by the constant use of water-scooping machines. There are so many activities and buildings that it is a mistake to call the Po delta 'natural'; almost nothing is the spontaneous work of nature.

It is clear that this situation and these cultural meanings are very distant from ecologist's cultures and from the crisis of values of modernity. The delta farmers, more than those in the 'highlands' (reclaimed in the past centuries), consider themselves as the protagonists of redemption from poverty and a hostile environment. It is difficult to change this mind-set and to call for a new balance with the natural ecosystem. The latter has only just been won and it is still threatening. These farmers must be sure that their land is hydraulically stable.

With this mix of cultural ideas and the need for security, the farmer sees the park as an enemy. He/she perceives it as a negation of his/her identity as a land reclaimer and land reformer. Probably the greatest cultural separation between the agricultural world and environmentalism lies precisely in the delta area.[5] Artificially and geometrically drawn fields, with no vegetation except for crops, are the pride of rational agriculture. This method allowed exploitation of every strip of land wrested from the water. In contrast, for an urban environmentalist culture, this is the negation of every attractive aspect of countryside charm, as well as being ecologically very poor land.

Beyond aesthetic, ecological and hydrological questions, the park perhaps signals the profound crisis of trust in the institutions and in the ideas that legitimize them. It is said that social redemption cannot be the ideal on which different social groups converge, it cannot be the point at which local society and national society meet. If this ideal is in crisis, a crisis in political representation easily ensues.

The park represents a new cultural model for the local population. In its more advanced version, it is conceived as a new Utopia, an attempt to combine development and protection of the environment. It is an ideal that goes beyond the ideal of social redemption. The former includes the latter. New postmaterialistic values are posed: not only equity and health but also social participation and harmony with the environment.

This new model is not well delineated and is certainly confusing for the people of the delta. They probably see it as opposed to their ideal of redemption. They certainly see it as the fruit of urban culture, a culture ignorant of local problems and only concerned about green spaces for town-dwellers. The politicians are seen as uncertain about which culture to represent or, worse, as cynically ambiguous. Indeed, Italian politicians have never been clear in their attitudes to park policy. They have tried to maintain a precarious balance between the urban and rural populations' expectations.

The major political difference between the Venetian and Emilian bank is the ability of the Left Democratic Party (former Italian Communist Party) to focus interests, to mediate conflicts and to impose its own decisions. In fact, this party has been able to posit the park proposal as a global policy in the Emilia-Romagna delta. Certainly, the mediation has had its price, a low-profile park (Poggi, 1995), but the conditions have been extremely difficult: first, long-drawn-out mediation among all social groups, then implementation of the regional law and finally the decision to proceed without the Veneto region and despite the hostility of the Minister of Environment – doubtless, not ready to give up the idea of a national park. The prestige of the Partito Democratico della Sinistra (Left Democratic Party) is so high that the Venetian delta people also look to the other side as a point of reference for their own action.

The reasons for the strength and stability of Emilia-Romagna Left Democratic Party are many. The most immediate reason is good local administration, or because this party, never having been in the government at the national level, has preserved an image of moral purity. Another reason is its ability, particularly in Emilia-Romagna, to combine ideological flexibility with internal unity. But, beyond these reasons, does the party still represent local civil society as a common ideal for the majority of the population? Or is its success over the delta park proposal linked to residual capacities, to traditional prestige? Is the party itself a victim of the slow erosion of the redemption ideal and will its loss of consensus emerge later? It is not easy to give an answer. Certainly, the new park will be a good test for this party and the other social groups not necessarily congruent with it (for example, farmers) to show whether they have a new cultural synthesis on which to found civil society.

CONCLUSIONS

Everywhere in Italy, park projects are in conflict with the agricultural world. The exception is the mountain parks, where agriculture is quite weak. In these areas, plurifunctionality and integration with other activities are seen by farmers as the only way to survive. In the lowland areas, where agriculture is integrated in *filières*, there are always farmers opposed to park proposals (Tempesta, 1994).

The establishment of a park has, then, exacerbated the agriculture–society conflict. It has triggered an identification crisis between sectoral interests and general interests. The different evolution of the park on the two sides of the River Po means that some attempts have been made to find a new synthesis between agriculture and society. How can this attempt be assessed?

The almost total exclusion of farms from the park border is only an apparent victory (a Pyrrhic victory). This establishes the separation between agricultural policies and territorial policies. The decision to wait and perhaps to enter the park area later is a signal of weakness. It means the farmers do not trust public bodies. It means they are unwilling to take the risk of institutional innovations. It means they have no clear idea about the trend in park and environmental policies. They probably reduce everything to an exchange: money to offset the income lost on ecoproductions. They are loath to assume risk in global policies for park land.

The legitimation of the agricultural world as the crucial agent of rural development, according to the new EU philosophy (Commissione delle Comunità Europee, 1988), fails to find an instructive example. In the Venetian part of the Po Delta Park, there is open hostility towards outside society, while in the Emilian part the logic of separation has prevailed. The new development philosophy, instead, is to integrate economic activities with environmental protection.

At stake, in fact, is environmental protection, as underlined by the criteria governing access to Objective 5b funds. The park, though basically flawed (being only regulatory), is conceived as a dynamic instrument with which to merge different interests (Parker, 1990). It may provide an opportunity to experiment with new types of environment-friendly production, to qualify goods with a mark, to start up agrotourism and obtain new resources.[6]

It requires, however, a major innovative effort and long-term trust. The first results will probably be forthcoming many years after the park has been created. A long wait entails trust in politicians, the abandonment of a short-term exchange perspective and the clarity of future trends. To date, the Po Delta Park has been an opportunity missed to re-establish the dialogue between agriculture and society. That, however, must not be imputed solely to the agricultural world. The local institutions were also not able to instil trust. This is seen, particularly, on the Venetian side, where the encircling syndrome has prevailed. In conclusion, the perspective that the agricultural world will be the pivot of rural development is very far away.

At this point, two questions arise. What will be the agricultural world's role in rural society? Which actors will try to conciliate interests in the name of environmental protection? In the former case, the agricultural world will probably persist with its sectoral policy, seeking to limit damage as much as possible. If this strategy is adopted, the isolation of farmers and their conflicts with other social groups will grow. In the latter case, the answer is more difficult. To the south of the River Po, there is a political party still able to focus on different interests and to act as a point of reference for public choices. But, if its capacity is only residual, an environmental management problem already dramatically evident in Italy will arise. Institutional arrangements are not enough: giving a place to every union on the park council is only a short-term solution. An answer must be found at the cultural level as well. A new shared mission, such as that of social redemption, must be sought. The farmers can still draw on their deep cultural heritage to help this search.

NOTES

1 When unions have the monopoly of farmers' representation, they can, according to public-choice theory, act as if 'rent seeking' in the political arena (Mueller, 1989, pp. 229–246).

2 Pesticide regulation is a case in point. In 1990, it was subjected to a national referendum, promoted by environmental associations. Other striking examples of corporatist regulation problems in Italy have been the financial crisis of Federconsorzi – the biggest second-degree cooperative in Italian agriculture – and the problem of milk-quota redistribution (Fanfani, 1993).

3 The crisis of the corporatist model goes beyond the agricultural sector and concerns the governability of society as a whole. A typical response to the crisis was the regionalism of both agricultural and environmental policies (Frows and Tatenhove, 1993) and of general institutional arrangements (Dente, 1985).

4 The political situation in the two delta subareas is very different. The delta itself is traditionally 'red' because of the high number of agricultural workers (*braccianti*). However, the northern part – in Veneto – is part of a politically less left-wing province and of a region in which the Christian Democrat Party has an absolute majority. In recent years, major changes have taken place: in the Venetian delta mixed coalitions have arisen, while at the regional level the Christian Democrat Party has rapidly lost its majority, provoking a long period of political instability.

The southern part of the delta – in Emilia-Romagna – has been dominated by the Communist Party at every level – municipal, provincial and regional – since the end of the Second World War. The Emilia-Romagna region was not subject to instability when large-scale corruption was discovered among politicians. But the judicial investigation provoked – particularly in Veneto – the collapse of the traditional parties and the formation of mixed and short-lived coalitions.

5 The Venetian part of the delta is younger than the Emilian part. The Venetian side comprises almost all the living river branches. This also helps to explain the different attitudes of farmers on the two sides.

6 Regulation 2078/92 provides grants for agri-environmental measures. These grants are larger in protected areas. The national framework law on protected areas sets priorities in granting public funds, particularly for farms located inside the parks (Agostino, 1993; Castiglione, 1994; Montresor, 1994). To date, these policies have not worked well in already-established parks. This has increased farmers' fears that inclusion in the park will negatively affect their income and land value.

4

The Values of the Agricultural Landscape: a Discussion on Value-related Terms in Natural and Social Sciences and the Implications for the Contingent Valuation Method

Fredrik Holstein[1]

INTRODUCTION

The agricultural landscape of Sweden is changing. In the middle of the nineteenth century, there were about 2 million ha of seminatural grazing land; today about 300,000 ha remain. Of the 2 million ha of meadows present at the beginning of the nineteenth century, 2500 ha remain. The threat to such landscapes is also a threat to the biological, cultural-historical, recreational, economic and aesthetic values attached to them (Jordbruksverket, 1994, 1995). The threat to these values can motivate policy measures, which, in turn, can motivate the estimation of the values. To formulate relevant policy goals, the values must be expressed and compared with each other and with other values, in some acceptable way, which, as a start, requires that the significance of the terms is understood by all actors.

An analysis where value-related terms are compared with value-theoretical concepts shows, however, that the meanings of the terms are unclear. Hence, there is a risk of confusion, resulting in less suitable policy measures, and thus a need for more precise terminology for describing the values of the landscape.

The first objective of this chapter is to analyse some value-related terms from a value-theoretical view. How can the different value-related terms be interpreted? Should economic, biological, cultural-historical, recreational and aesthetic values of a landscape be added to obtain the total value of the landscape? Or do the different terms indicate different perspectives of value which might not be compatible? Brown (1984) states that 'only when economic value measures are placed in perspective do their nature and shortcomings come

 The Economics of Landscape and Wildlife Conservation (eds S. Dabbert, A. Dubgaard, L. Slangen and M. Whitby)

clearly to light'. This is true also for other concepts of value. The intention of this chapter is to broaden the discussion and to make interdisciplinary discussions easier. This should lead to better communication between scientists representing different disciplines, politicians, administrators, farmers and the public. Better communication increases the possibilities for better policy instruments and more of the values, however we choose to define them, of the agricultural landscape can be preserved.

The second objective of the chapter is to discuss the complications of different perspectives of values when applying the contingent valuation method (CVM). The analysis of CVM-related problems is concentrated on the problems of different views of values, but also the problems connected with the distribution of property rights among the respondents is briefly discussed. Other problems, such as the choice of questioning modes and econometric techniques when applying CVM, are not discussed.[2]

WHAT IS A VALUE?

The preservation of or changes in landscapes can be motivated by the values attached to them. In fact, as Lockwood (1997) states, 'value assessment is a necessary component of any rational decision-making process concerning the use or management of natural areas'. By different use of the landscape, different landscape-related values will be realized and, since every landscape-management act requires resources, there is always an opportunity cost – that is the resources used for managing the landscape could be used to realize other values. So the question of values is crucial for any rational decision about how the landscape should be used. But what do we mean by the term 'values'? The question of values has been discussed by, among others, philosophers and economists for hundreds of years and it is not surprising that the term has many meanings (Brown, 1984).

Different attempts to describe values have been made. Brown (1984) distinguishes three uses of the term. The first meaning is the purely mathematical, as in the expression 'the value of a variable'. The second meaning concerns physical or biological relationships and can be described as 'functional value relationships' (Andrews and Waites, 1978, cited in Brown, 1984). The third meaning, which Brown is concerned with, is 'the preference-related concept of value'. Brown also distinguishes between held values and assigned values: 'the preference relationship between a person and an object, given the person's held values, results in different objects' assigned value'. The economic measures of value are a set of assigned values. Lockwood (1997) distinguishes between intrinsic, functional and instrumental values. Intrinsic values are interpreted as end values. The term functional values has the same meaning as Brown's 'functional value relationships'. If this functional relationship contributes to the realization of some end value, the process or the entity will have an instrumental value.

Both Brown (1984) and Lockwood (1997) are concerned with human values. However, there are other ways of interpreting values, in the sense that something is better than something else. Lockwood admits the possibility that species other than humans have end values and that these, if that is the case, should be included in an extended ecological economic analysis.

In the following, some value theoretical concepts will be presented. The concepts are mainly based on Ariansen's 'environmental philosophy' (Ariansen, 1993). The concepts will thereafter be used to analyse some value-related terms often used to describe the landscape and to motivate policy instruments toward the landscape.

Value Statements and Descriptions

First, the distinction between a value statement and a description should be clarified. It was pointed out by David Hume that it is not possible to make value statements from pure descriptions. Any value statements made without a value premiss are said to be a 'naturalistic fallacy' (Ariansen, 1993).

Assume that there will be some degradation in biodiversity if the current 300,000 ha of seminatural grazing lands and meadows are not preserved in Sweden. From this description, we cannot say if the current areas should be preserved. To be able to draw a conclusion like that, we need a value premiss, such as 'all biodiversity should be preserved'. The value premiss is an indicator of the viewpoint; in this case, the landscape might be seen as a part of the ecosystem. From the value premiss 'all biodiversity must be preserved' and a (here assumed) knowledge that there will be losses in biodiversity, one can, without any logical mistakes, conclude that the current 300,000 ha of seminatural grazing landscape should be preserved.

It should be mentioned that the distinction made above is rather problematic. Are there any pure descriptions, neutral to valuations? It can be argued that the way we observe and describe the world depends on our valuations. Even the language might be constructed from valuations; we have chosen to construct words for those phenomena that concern us in some way. The fact that there is a word for biodiversity indicates that we are (or at least someone is) interested in biodiversity: the phenomenon is relevant. It does not, however, tell us whether biodiversity is good or bad. Even if the distinction is not obvious it will, even in the following, be assumed that it is possible to distinguish 'is' from 'ought'.

Due to this distinction, the term 'functional value relationships' is not a value statement at all but rather a description.

Value Objectivism and Value Subjectivism

The next question concerns the nature of values. Are values a part of the objects or are they in 'the eye of the beholder'? The question indicates two different perspectives on values: the value-subjective and the value-objective perspectives. According to the value-subjective perspective, values are 'created' by the subject. This is the perspective of, for example, Brown (1984), when he states that 'value is neither a concept held by the subject nor something attributed to the object, but merely that which arises from the preference of a subject for an object'. However, the relation between the subject and the object does not necessarily have to be in the form of preferences. According to the value-objective perspective, values are properties of the objects, just like length and weight. In the view of the philosopher Holmes Rolston III, for example, there are objective intrinsic values. These values can be seen as potentials, which can be realized by someone experiencing them (Anderberg, 1994). Here, the term intrinsic value is used to describe the value-objective perspective.[3]

The value-subjective perspective is often combined with the view that humans are the only valuing subjects.[4] However, here it is stressed that these are two different questions and it is possible to think of subjects other than human beings.

When it comes to human decision-making, the perspective of values defines the role for humans. The conclusion is that there are three possible roles for humans:

1. As the only relevant valuing subjects.
2. As valuing subjects and as interpreters of, for example, other species' valuations.
3. As interpreters of objective intrinsic values.

Roles 1 and 2 follow from different value-subjective perspectives and role 3 follows the value-objective perspective. The traditional welfare-economic analyses assume role 1 for humans. This means that it will be difficult to accept the results of a welfare-economic analysis, or any other analysis with an anthropocentric value-subjective perspective, if one prefers role 2 or 3 for humans.

Instrumental Values and End Values

No matter how we have defined values, as objective or subjective, something can be valuable because it serves some other value; it then has an instrumental value. The other value, realized by the instrumental value, might also be an instrumental value. This chain of instrumental values can be long, but not infinite. Somewhere there must be an end and those values are here called end values. An end value can either be motivated as a subjective value or as an objective intrinsic value. If every species has an intrinsic value and if the agricultural landscape is crucial for the survival of at least one species, the landscape has an instrumental value in relation to the end value of that species (Ariansen, 1993; Lockwood, 1997).

Subject Motives, Groups of Objects and Functions

A hypothesis of this chapter is that many of the value-related terms used for motivating policy measures toward the landscape have an unclear connection with the theory of values. It would probably be difficult or impossible to relate them to the value theory presented above. A further classification is therefore needed. The one presented here follows the logic of Hasund (1991), with small modifications of the terms.

Subject motives for valuation describe the motives of the valuing subjects. The value of a landscape can (adopting the anthropocentric value-subjective perspective) be motivated by the possibility of using the landscape (for example, recreation) or by the subject enjoying the pure existence of the landscape. Using the terminology proposed here, the values are user-motivated values and existence-motivated values, respectively.[5]

Groups of objects of value refer to a class of objects. The objects of this group might have value motivated from the value-subjective perspective or from the value-objective perspective.

Functions refer to the relationship between phenomena and the term is used here with the same meaning as 'functional value relationships', as used by Brown (1984). In the cases where an object has a function for a subject, this is described as subject motives. Hence, the term functions is used to describe relationships between objects. Since those relationships do not have to indicate any value statement, the term value is avoided. Those functions can realize values either directly or indirectly. A specific species can, for example, have a function in an ecosystem. If that species is essential for the survival of other species that are of value, it will have an instrumental value.

VALUE-RELATED TERMS AND THE THEORY OF VALUES

In the introduction, some value-related terms were mentioned. Those terms – biological, cultural-historical, recreational, economic and aesthetic values – are analysed here, using the terminology presented in the previous section. The question is whether the terms indicate any specific perspectives of value or if they have to be interpreted from different perspectives. Moreover, is it possible to understand them from a common perspective, for example from a welfare-economic perspective?

Aesthetic, Recreational, Cultural-historical and Biological Values

The term aesthetic value does not indicate if the perspective is value-objective or value-subjective. The term can be interpreted from the value-objective perspective if there is such a thing as objective beauty. To clarify the perspective in that case, the term intrinsic aesthetic value can be used. Another interpretation, from the value-subjective perspective, is that the term aesthetic value describes a subject motive, for valuing, for example, a meadow, and hence describes an aesthetic motivation for the valuation.

The term recreational values must be understood from the value-subjective perspective; it is hard to think of any intrinsic recreational values. However, the term seems to refer to a subject motive, a user-motivated valuation. The term recreation-motivated valuation would indicate the perspective.

The term cultural-historical value also seems to have weak connections with the value-theoretical concepts presented above. The most reasonable interpretation is that the term describes a group of objects, objects regarded as cultural-historic in some sense. A cultural-historic object does not necessarily have to be of value. The fact that someone pays attention to that object will often indicate that he or she cares about it and the value statement can be implied. However, a more stringent description would be to use the term cultural-historical object of value.

The term biological values can be interpreted from quite different perspectives; hence it is likely that misunderstandings will occur if the intended meaning of the term is not clarified. One interpretation is that the term refers to a group of objects, biological objects of value. In this case, the values can be value-subjective as well as value-objective; the term does not imply any specific value-theoretical perspective.

The term can also refer to biological functions of value. If a species has a crucial function in an ecosystem, which has a value (for example, user-motivated), that species will have an instrumental value. This illustrates why knowledge about ecological functions is essential for valuations of, for example, meadows, even if an anthropocentric value-subjective perspective is adopted.

The term might also be used to describe the value of, for example, herbs indicating the presence of other, for some reason, valuable species or habitats. In this case, it seems more stringent to use the term function. The herb in question has an informational function. This function will have a value, an instrumental value, if the use of the information realizes some other value.

Finally, the term can be used from the value-objective perspective if it is assumed that all biological life has an intrinsic value. To clarify this perspective, the term intrinsic biological value can be used.

Economic Value

The term economic value seems to have a rather clear meaning in relation to the value-theoretical concepts. The perspective is value-subjective and the only relevant valuing subjects are humans. Values are derived from human preferences and can be approximated by consumer and producer surpluses. The total economic value (TEV) is understood as the aggregate of all user-motivated and non-user-motivated values (Pearce and Turner, 1990).

It has been argued, by Gren *et al.* (1994), that the concept of TEV does not capture the total value of the ecosystems. Gren *et al.* (1994) use the terms primary and secondary values, where primary values refer to the self-organizing capacity of the ecosystems. Those primary values are not, according to the authors, captured in the TEV. Secondary values refer to the production of goods and services. In the view of values adopted in this chapter, those terms seem to be misleading, since the distinction is made without any clear connection to the theory of values. It is suggested here that the terms primary functions and secondary functions are more useful in this context. From the welfare-economic perspective, the primary functions will then have values, recognized either as instrumental values (if those functions are necessary for any secondary functions appreciated by humans) or as end values (if someone finds an existence-motivated value in the functions themselves).

As long as the welfare-economic perspective is adopted, the concept of TEV captures all values. But that is a different question: the total value will probably never be elicited in practice. If the viewpoint is changed (to, for example, non-anthropocentric and/or value-subjective), other values will be found, but at the same time the concept of TEV will lose its meaning. So, if the values of the agricultural landscape are described from two (or more) viewpoints, those values cannot be added together! The different analyses should rather be seen as different pictures of the same reality: the viewpoints are complementary but the resulting values are not!

In debate, economic value is often interpreted as a value to be summed up to other values. This seems to be the case in some publications from the Swedish Board of Agriculture (Jordbruksverket, 1994, 1995), where different values are described as if they should be added together. One possible explanation is that economic value is interpreted as market price or the profit of the farmer. Jones (1993) distinguishes three types of economic value, one of which is market value.[6] In his discussion about different types of values associated with the landscape, it is clear that economic values (market values) are not interpreted as a concept for describing the total value. The term TEV does not seem to solve this problem. Maybe the term economic measure of total value (EMTV) would be more informative? This term emphasizes, contrary to TEV, that it is the total value and, hence, that nothing more can be added.

Value-related Terms from the Welfare-economic Perspective

In the previous section, some value-related terms were interpreted from the value-theoretical perspective presented earlier. It was concluded that many of the terms have rather unclear connections with the theory of values. However, welfare economics offers one possible way of describing values. If this perspective is chosen, is there still a meaning in the other value-related terms or do they have to be interpreted from other perspectives? The analysis above has shown that there are possibilities for interpreting them in a way which should be compatible with the welfare-economic perspective. In this section, the values of landscape are described from the perspective of welfare economics and the other value-related terms are interpreted so that they fit into that perspective. As should be clear from the above, this does not mean that these interpretations are the only correct ones.

Figure 4.1 illustrates an agricultural landscape, for example a seminatural grazing landscape. The value premiss is the satisfaction of human preferences. How can the values of the landscape be described if we adopt the welfare-economic perspective?

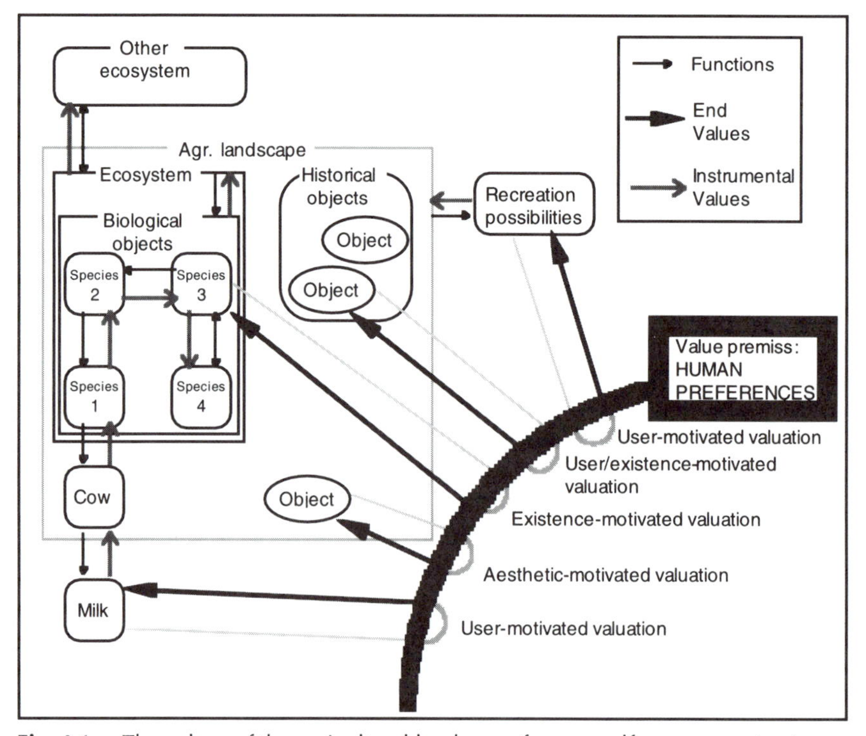

Fig. 4.1. The values of the agricultural landscape from a welfare-economic view.

To start with the value of species, it can be seen that the very same species can have an instrumental value and an end value. This is illustrated by species number 3 in Fig. 4.1. Species number 3 has some kind of biological function for species number 2, which, in turn, has some biological function for species number 1. This later species is grazed by cows, cows producing milk for consumption. The fact that humans consume the milk indicates that there are preferences for milk. Human preferences are, in this case, our value premiss and hence the milk has an end value. The cow has, then, an instrumental value and instrumental values can be derived back to, in turn, species 1, 2, 3 and 4. At the same time, there may be, as in this example, someone with preferences for the pure existence of species number 3, i.e. it has an existence-motivated end value. We can conclude that there are biological objects of value in the landscape.

There are also other objects of value in the landscape. There is a cultural-historical object of value, where the valuation can be user- and/or existence-motivated. Another object has a value due to an aesthetic-motivated valuation.

Can it be claimed that the agricultural ecosystem has a value? If we assume that there are primary functions in the ecosystem in the agricultural landscape and that these functions are essential for the survival of the species (the biological objects of value), there is value in the ecosystem itself, an instrumental value. We can also talk about biological functions of value. As illustrated in the top left corner, there might be instrumental values assigned to other ecosystems as well, motivated by their functions for the agricultural ecosystem.

The landscape as a totality can offer recreation possibilities. Hence the agricultural landscape can have a user-motivated value.

It has been shown that welfare economics offers a possibility for defining the total value of an agricultural landscape. Moreover, it is possible to interpret other value-related terms so that they have a meaning in that perspective. However, from this it cannot be concluded that the welfare-economic perspective of values should be preferred to any other perspective. Such a conclusion has to be based on meta-ethical principles, making it possible to rank, for example, perspectives of values. Such a ranking can probably never be done and the problem can be handled by a careful use of value-related terms together with conditional conclusions. It is the responsibility of the economist to be aware of and discuss the approach used, including the ethical basis for the analysis. This might be sufficient, for example, when advising politicians but the question of different perspectives of values becomes more crucial when it comes to economic methods, such as CVM, where respondents are supposed to answer questions from a specific perspective of values.

THE CONTINGENT VALUATION METHOD AND DIFFERENT PERSPECTIVES OF VALUES

The general welfare-economic context can be seen as one picture of reality. The CVM is one method of eliciting values through the measure of willingness to pay (WTP) or willingness to accept (WTA). There is an ongoing debate on the validity of CVM, where some of the problems referred to are embedding or the problem of scope, strategic bias, hypothetical bias, design bias, instrument bias, starting-point bias, warm-glow effect, non-response bias and protest bias (Jakobsson and Dragun, 1996). Sagoff (1994) has pointed out that CVM requires that the respondents must adopt the same idea of values that underlies the method. Otherwise, there is a risk that the answers will be directed towards a question other than that intended by the survey constructor and the interpretation of the results will be quite problematic (Sagoff, 1994). The main focus here is to discuss whether some of the problems of CVM can be explained by respondents having different views of values. This explanation does not exclude other explanations; different views of values might, for example, result in protest bids.

Above, it was concluded that welfare economics requires an anthropocentric value-subjective perspective. This perspective does not, however, automatically lead to the welfare-economic perspective. Sagoff (1988, 1994), for example, stresses that private preferences are not the relevant measurement of values and that 'an analyst who asks how much citizens would pay to satisfy opinions that they advocate through political association commits a category mistake' (Sagoff, 1988, p. 94). Even if the analyst does not agree with the distinction between private and public preferences, the respondents may very well do so and there is a substantial risk of protest bias.

If the welfare-economic perspective is still chosen, that is, humans are the only valuing subjects and values are measured through preferences, there are more questions that must be answered before the analyses can be carried out. The question of which measure of welfare change to use, WTA or WTP, has been discussed extensively in the literature. It has been shown by Hanemann (1994b) that there is no theoretical presumption that the measures will be close in value. The recommendation from the National Oceanic and Atmospheric Administration (NOAA) Panel to always use the WTP format is motivated by the panel's desire to obtain a 'conservative' result from the contingent valuation (CV) studies (Arrow *et al.*, 1993). It is, however, dangerous, as pointed out by Harrison (1993), to measure one thing (WTP) if the correct thing to measure is something else (WTA), unless you know, in advance, that the measures will give approximately the same answers. Thus, there are no theoretical arguments for using the WTP format in situations when the property rights are distributed in such a way, that the respondent has a right to the amenity in question, making WTA the correct format (Jakobsson and Dragun, 1996).

Contingent Valuation Method Respondents' Views of Values

A study carried out by Vadnjal and O'Connor (1994) was 'designed to obtain information on how people actually interpret questions of paying'. In the study, respondents in the East Auckland suburb of St Heliers Bay, New Zealand, were asked to express their views on the 'value' of Rangitoto Island. The island is now undeveloped and the respondents were shown two pictures of the island: one of the island in its present state (View A) and another with extensive building on the island (View B). In addition to the hypothetical WTP question,[7] respondents were asked which view of the island they preferred. As a follow-up question, the respondents were asked:

> Do you think the money amount you finally agreed to pay is an accurate measure of the value to you of continuing protection of View A?

Those who did not think so were asked to explain why they did not think that their stated WTP was an accurate measure of the value. The conclusion from the answers was that 63% of the respondents stated verbal interpretations at odds with the interpretation of values in welfare economics and CVM. A further sequence of questions was asked to find out more about the respondents' views on 'environmental values' and Rangitoto Island. The authors conclude that, even if the majority of respondents state a WTP, it is quite probable that those figures will have no meaning in the context of CVM.

> They responded in terms of a view of how things ought to be in society, not from the standpoint of a merely self-interested consumer. This was made clear in two main ways: the widely expressed view that one 'shouldn't have to pay', and the sentiment also widespread that payment signified a willingness to 'fight to keep it'.
>
> Vadnjal and O'Connor (1994, pp. 377–378)

It was further concluded that people want to find something in their environment that is beyond choice and value. In the economic context, one would stress that there is nothing beyond choice: to place something beyond choice is also a choice. If we move away from the economic viewpoint, this might still be 'true' but it is not obvious from which value criteria these choices should be made: is it a question of weighing preferences, a question of what is right or wrong or another kind of question?

When people expressed views of how things ought to be in society, this could be interpreted as another view of property rights, which would make the WTA format more relevant. However, the respondents in the survey about Rangitoto Island were asked if something could compensate for the loss of the unexplored island and the answer was that nothing could compensate; it would, as some expressed it, be a shame. This could, if one insists on the welfare-economic viewpoint, still be interpreted as a question of property rights. In cases concerning common goods it is possible that the respondents refuse to state their WTA as long as all the other owners (including future generations) have not agreed to selling the property. Then the question should be put, 'If everyone else agrees on the development shown in View B, what is the minimum amount of money you would be willing to accept to agree?' But, if the respondents still do not adopt the viewpoint of welfare economics, there will be a problem. Using the WTA format would solve some of the problems (and maybe raise new ones), but that solution would not be sufficient if, for example, some of the respondents believe that species other than humans are also relevant valuing subjects or that there are intrinsic values.

How does it occur, then, that respondents state a positive WTP even if they do not think that they should have to pay for 'something that's already yours'? According to Vadnjal and O'Connor (1994), the WTP can be interpreted as gestures in a political process. Many said that they would contribute money for a 'fighting fund to make sure nothing changes'. So the interpretation is that there is a WTP, not for Rangitoto Island but rather for one's right to participate in the political process, a right which most of us think should not be motivated by a WTP at all. One conclusion might be that respondents are answering more as citizens, making judgements of what is right and wrong, than as consumers. This is also the conclusion of Blamey *et al.* (1995), referring to a study of forest management in Australia undertaken by the Resource Assessment Commission (RAC).

Different Views Explained as Lexicographic Preferences

The problem of how respondents perceive the CVM questions is also investigated by Spash and Hanley (1995). In a survey, they have investigated the WTP to prevent a hypothetical reduction in biodiversity. Using the results, they investigate the nature of preferences, coming to the conclusion that CVM fails in measuring welfare changes due to the prevalence of lexicographic preferences.

Lexicographic preferences describe a case where the individual refuses to make any trade-offs to compensate for the loss of a particular commodity. Using the welfare-economic measures, WTP would be equal to the entire budget and WTA would be infinite.

In Fig. 4.2 the prevalence of lexicographic preferences for W (for example, wildlife) is illustrated. Assuming two goods providing utility and W having a priority over the other good, the 'indifference curves' are reduced to 'indifference points'. Starting at point A there is no increase in X that is enough to compensate for a loss in W. Since W has priority over X, all points in the shaded area (including the line segment AB) are preferred to A. That is,

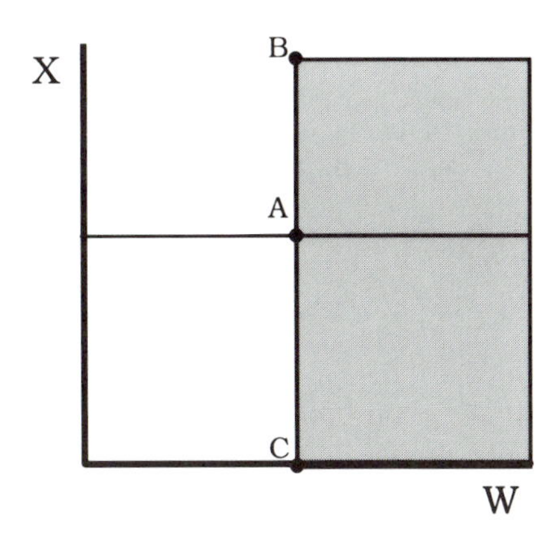

Fig. 4.2. Lexicographic preferences (from Spash and Hanley, 1995).

there is no point producing the same utility level as A, or no indifference curve exists.

The implication of lexicographic preferences is that one good (X) will have no priority at all, illustrated as point C being preferred to all points to the left of the line BC. In the example used by Spash and Hanley (1995), this means that all other consumption will be sacrificed to keep the wildlife at the level of the line BAC. This results in the existence of lexicographic preferences being regarded as unrealistic.[8] Spash and Hanley conclude, before going on with some empirical evidence, that, if there are lexicographic preferences involved, the Kaldor–Hicks potential compensation test will become inoperable. This seems to be a rather strange conclusion: why is a test inoperable because it sometimes gives the answer 'no, this project should not be carried out'? In the theoretical example (illustrated in Fig. 4.2) used by Spash and Hanley (1995), they suggest that the motivation for not accepting any compensation might be that an individual believes that 'aspects of the environment, such as wildlife, have an absolute right to be protected'. Without saying anything about the possible existence of lexicographic preferences, this seems to be an example where it simply is not about preferences at all! It is rather about the role of humans and the argument can be interpreted as supporting the standpoint either that humans have to interpret other species' valuations or that they are intrinsic. If someone else has a right, why should I answer a question about my WTP for preserving the 'property rights' that already exist? What is the meaning of my WTA compensation for the selling of what belongs to someone else? The motives that Spash and Hanley (1995) refer to should not be interpreted as 'lexicographic preferences', not as preferences at all. If respondents report WTP or WTA in a CV study, this should be interpreted in the way proposed by Vadnjal and O'Connor (1994).

The Contingent Valuation Method as Surrogate Referendums

Blamey *et al.* (1995) discuss the possibility of using discrete choice (DC) methods (or referendum methods) to avoid some of the problems discussed above. They conclude that the result of such studies should not be interpreted as welfare measures but that 'referendum votes are more likely to represent statements about the socially optimal level of provision of particular public goods than they are to yield estimates of individual benefit' (Blamey *et al.*, 1995, pp. 268–269). They also present empirical results supporting the idea that the respondents act as citizens rather than as consumers when answering CVM surveys. Their conclusion is that CVM studies do not produce results appropriate for use in welfare-economic analyses, such as cost–benefit analyses (CBA). Their results, however, increase the appeal of the use of CVM studies as 'surrogate referendums'.

If humans are to interpret other species' valuations or intrinsic values, this is probably handled better in democratic discussions than through a CV study designed to elicit welfare-economic measures of value. Also the referendum format seems to be more attractive.

CONCLUSIONS

This chapter has discussed the concepts of value and it can be concluded that there are different views of values, where welfare economics represents one of these views. Other value-related terms, such as biological values, can be interpreted from different perspectives of value. Hence, those terms could be understood through and incorporated in a welfare-economic analysis only if they are interpreted from an anthropocentric value-subjective and preference-based perspective of values. The meaning of value-related terms would be clearer if authors indicated the adopted perspective of values. Biological values, for example, can be interpreted from different perspectives and this would be clarified if terms like biological functions of value, biological objects of value and intrinsic biological value were used. The meaning of economic value would be clearer if the term EMTV were used.

The term values can be differently understood, depending on the viewpoint. It is most probable that people have different world-views, including different interpretations of this term. Although it may be reasonable to accept the welfare-economics approach as one way of interpreting these terms, the divergence in the understanding of them is a problem for the CVM. If the respondent does not share the value-subjective and anthropocentric view of welfare economics, the interpretation of the WTP/WTA question might be quite different from that intended by the researcher. There is a risk, in the CBA context, that the stated WTP/WTA expresses views on, for example, the rights of species or ecosystems and not the preferences.

Lexicographic preferences are not the problem. If there are any lexicographic preferences, this is not to be regarded as a problem in the CBA context. The method should not be rejected by the argument that it sometimes gives the answer that a project should not be carried out. If there really are any lexicographic preferences, the distribution of property rights becomes crucial for the outcome of the CVM analysis, as well as for the CBA. But some of the problems underlying the use of the term lexicographic preferences are important to the use of CVM. The conclusion is, however, that these problems should be described as differences in views on values and how property rights should be distributed, and not as preferences at all.

The question of whether the species or ecosystems have any rights is particularly interesting when discussing the agricultural landscape. If there are any rights to a preserved landscape (held by the consumer and/or the species or ecosystem), there must be an obligation for someone (for example, the farmer) to act. This is quite different from the situation where the obligation is 'not to exploit', as in the case with Rangitoto Island. If it is an obligation for humanity to support the species dependent on a specific kind of agricultural landscape, it might be the case that the public should compensate the farmers for maintaining the responsibility of the human society. Whether there is any use for CVM if this view is adopted remains to be discussed. It could, however, be the case that the obligation follows, or should follow, with the ownership of the land. In a CVM study concerning the agricultural landscape, these questions have to be resolved.

So what should be learned by the survey designer? Either one must force the respondents to adopt the welfare-economic view and the proposed definition of property rights or one must be very careful with the interpretation of the stated WTP/WTA. In all cases, the usual WTP/WTA question should be complemented by questions about the respondents' views on values, ethical considerations and property rights. By discussing different views of the problem, it might be easier to find acceptance for using the welfare-economic perspective as one perspective among others. If the respondents are allowed to discuss and state their views on values before the WTP/WTA question is asked, it might be easier to 'hypothetically accept' the view and answer the question intended by the survey designer. However, it is also possible that a discussion of these questions will be confusing and that the interpretation of the answers will be even harder. As always, focus groups and prestudies are probably of great value.

NOTES

1 I would like to thank Thomas Hahn, Knut Per Hasund and Dan Vadnjal for helpful comments. The author is, however, solely responsible for all errors and omissions.

2 For a discussion of practical problems in CVM, see Kriström (1995).

3 O'Neill (1992) distinguishes three interpretations of the term intrinsic value: (i) end value; (ii) value as an intrinsic property; and (iii) objective value. Here the term is used in the third interpretation.

4 This is the case in Ariansen (1993).

5 These terms are used here instead of the environmental-economic terms, well known in the literature, user value and existence value, introduced by Krutilla (1967).

6 The other two are subsistence value and utilitarian ecological value.

7 There were actually two questionnaires used, QI and QII. The scenario of QI was that approval for the building existed and the respondents were asked to state their maximum WTP into a private trust fund to avoid that development. In QII, it was assumed that the island was protected and the respondents were asked to state their WTP for continuing protection.

8 Malinvaud (1972, p. 20), referred to in Spash and Hanley (1995).

Economic Valuation of Recreational Benefits from Danish Forests

5

Alex Dubgaard

INTRODUCTION

In 1989, the Danish government initiated a national afforestation programme aimed at doubling the forest cover from the present 12% to about one-quarter of the country's total area. The time horizon for this plan is a tree generation, about 100 years. When the programme was conceived, it was expected that large areas of agricultural land would be removed from production and afforestation was seen mainly as a residual land-use option. Gradually, Danish afforestation policies have taken a more integrated approach, emphasizing multipurpose afforestation projects, where recreation plays a major role. This means that the social efficiency of the afforestation policy depends to a larger extent on the provision of non-market benefits, especially recreation opportunities, which in turn calls for the inclusion of recreation in economic appraisals of afforestation projects. The study reviewed here is a contingent valuation (CV) survey eliciting the general public's willingness to pay (WTP) for access to outdoor recreation in Danish forests. It is the first attempt in Denmark to estimate the recreational benefits from the entire forest area. The presentation of the CV results is followed by an *ad hoc* cost–benefit appraisal, showing the order of magnitude of recreational net benefits from Danish forests. To provide information about the policy setting, the chapter begins with an overview of the Danish afforestation programme and its historical background.

 The Economics of Landscape and Wildlife Conservation (eds S. Dabbert, A. Dubgaard, L. Slangen and M. Whitby)

DANISH AFFORESTATION POLICIES

With forests and woodlands covering about 12% of the land area, Denmark is one of the least afforested countries in Europe. The natural vegetation is (deciduous) forest, but most of the present forest cover is the result of afforestation efforts during the past two centuries. At the beginning of the nineteenth century, no more than 3–4% of the country could be classified as forest. Overexploitation and agricultural expansion were the main causes of forest decline. To halt the deforestation process, an ordinance was implemented in 1805 protecting remaining woodland from conversion to other uses. In the second half of the nineteenth century, the reclamation of the vast Danish heathlands gave momentum to reforestation efforts. Tree planting on seminatural areas (especially heathlands and dunes) continued during the first half of the twentieth century and, by 1950, forest cover had increased to about 10% of the country's total area. Since the beginning of the 1960s, afforestation has taken place mainly on marginal agricultural land (Jensen, 1976).

In the mid-1980s, it was widely expected that large areas of low-grade agricultural land in Denmark would be 'threatened' by the cessation of farming operations, due to deteriorating economic returns in crop production. This led to the formulation of the ambitious afforestation programme, aiming at doubling the forest area over the next 100 years (Dubgaard, 1996b). The more concrete objective is annual afforestation of 5000 ha of agricultural land. About half of the target area is intended to be afforested by the State. The marginalization predictions of the late 1980s have not materialized (Dubgaard, 1996b). This has contributed to changing the scope of the state afforestation scheme. The emphasis is now on multipurpose afforestation projects in urban fringe areas, where outdoor recreation and landscape amenity play a major role. However, the costs of land acquisitions in urban fringe areas and planting of amenity forests are considerably higher than afforestation of marginal agricultural land in remote areas. Increasing financial and social costs call for more scrutiny in afforestation policy appraisals. It is essential that public or supported afforestation projects pass a cost–benefit analysis incorporating the full social opportunity costs and benefits of the contemplated initiatives.

VALUATION OF WILLINGNESS TO PAY FOR FOREST RECREATION

Most of the benefits provided by outdoor recreation areas are either public or open-access goods. In principle, such forms of recreation as pleasure walking, bird watching, picking berries and so on are quasi-private goods. Property rights in these benefits can be defined and users excluded from access. However, in Denmark, legislation grants the right of common access to private as well as public forests. Consequently, as far as recreation is concerned, forests in Denmark are an open-access resource. This means that non-market valuation methods are required to assess the social value of the benefits from forest recreation. The present study includes (only) the on-site use value of access to outdoor recreation in Danish forests. In public forests, right of access includes non-consumptive use (pleasure walking, hiking and so on), as well as certain forms of consumptive use (picking of berries and mushrooms). In most private forests, only non-consumptive use is permitted.

Design of the Valuation Scenario

The recreational benefits from access to Danish forests were evaluated using the CV method. The CV technique attempts to elicit individuals' WTP for non-market goods where actual economic behaviour is not observable. To identify a demand curve for a non-market good, respondents are asked to state their WTP in a hypothetical market setting where the attributes of the good, the rules of provision and the method of payment are described (see, for example, Mitchell and Carson, 1989; Freeman, 1993). To be able to test the economic consistency in the WTP responses, information about respondent characteristics (such as socio-economic and demographic variables) is also collected. The final step is to obtain an aggregate benefit estimate, which is usually derived by multiplying sample mean WTP by the entire population of individuals benefiting from the provision of the good considered.

Survey instrument and sampling procedure

The present valuation study was integrated into the forest part of the project 'Outdoor Recreation 1995', conducted by the Danish Forest and Landscape Research Institute (FSL) (Jensen and Koch, 1997).[1] The FSL was responsible for sample selection and survey administration. The target population was all permanent residents in Denmark within the age groups 15–76 years. The Danish Civil Register was used as the sampling frame, which virtually precludes the occurrence of sampling-frame bias. The survey instrument was a mail questionnaire distributed to a national area probability sample of 2895 individuals during a 1-year period (1 November 1993 to 31 October 1994). An introductory letter explained the relevance of the survey for forest management and so on, and a self-addressed return envelope was enclosed.

Elicitation format

An open-ended (OE) CV question was used to elicit respondents' WTP for access to forest recreation (Box 5.1). The OE format invites respondents to freely state their (maximum) WTP – with no specific clues as to what the relevant level of payment might be. Asking an OE CV question is the simplest way to apply the

Box 5.1. Wording of CV question (version A).

In this survey we will also attempt to assess the economic value of recreation in Danish forests. Try to imagine that visitors had to buy a permit for access to forests.*

WHAT IS THE LARGEST AMOUNT YOU WOULD HAVE BEEN WILLING TO PAY FOR AN ANNUAL PASS GIVING YOU ACCESS TO USE DANISH FORESTS TO THE SAME EXTENT AS YOU DID LAST YEAR?

The highest amount I would have paid for an annual pass is:............ DKK

IF YOU HAVE ANSWERED 0 DKK, PLEASE INDICATE WHY:

I have not used forests for recreation during the last year.........

I have used forests for recreation, but I can't say what I would have been willing to pay for admission...........

I could not have afforded to pay.............

Something else:..

* There are no plans to charge the public for admission to recreational areas in Denmark.

CV method. The dichotomous choice (DC) approach is an alternative elicitation procedure, asking each respondent if he or she would pay a prespecified amount (bid) for access. This take-it-or-leave-it procedure has the conceptual advantage that it resembles the way consumers make choices in a real market and, according to the National Oceanic and Atmospheric Administration (NOAA) Panel,[2] it is the preferable approach from a methodological perspective. However, in the present study, the survey design (determined by the costs of administration and circulation of the questionnaire) only permitted the breakdown of the overall sample into four subsamples. This would limit the range of bids in a DC experiment to four different amounts. With no opportunity to pretest the design of an optimal bid structure, a bid range of this size would not have sufficed to facilitate subsequent estimation of WTP. Thus, given the survey design, the OE format was the only feasible approach.

Payment vehicle

Under the assumption that entry to Danish forests was restricted, the CV scenario constructed a hypothetical market for access to forest recreation in Denmark (Box 5.1). The payment vehicle was formulated as a lump sum for unlimited admission to all Danish forests for an individual during a 1-year period (annual pass).[3] Considering the existence of common access rights in Denmark, most respondents would undoubtedly find the introduction of user payment illegitimate. Those opposing a user fee might react by free-riding (understating WTP) or not responding to the WTP question (scenario rejection). To emphasize the hypothetical nature of the CV scenario, it was explicitly stated that 'There are no plans to charge the public for admission to recreational areas in Denmark.'

Protest bidding

In most CV studies, a significant number of the respondents will fail to give valid WTP amounts, either by not answering the WTP question or by stating implausibly low or high amounts (outliers). Usually, most of the zero bids are protest responses that do not reflect the true WTP of the respondent (Mitchell and Carson, 1989). From a methodological perspective, protest bidders should be removed from the sample, since a protest response is equivalent to rejecting the valuation scenario and would bias the estimates of mean and aggregate WTP.[4] To be able to identify protest responses, zero bidders were asked to specify the reason underlying the refusal to pay any positive amount (Box 5.1). Of the reasons suggested in the questionnaire only the following were considered 'legitimate' justifications for a zero bid: 'I have not used forests for recreation during the last year' or 'I could not have afforded to pay'.

Survey Results

A total of 2895 questionnaires were mailed to randomly selected residents in Denmark. Of those, 2424 questionnaires were returned, resulting in a highly satisfactory survey participation rate of 83.7%.[5] Only 159 of the individuals returning the questionnaire had failed to answer the CV question. This equals an item non-response rate of 6.6% (of the questionnaires returned).

Sample trimming

Close to 1000 respondents had stated a zero bid, about 40% of the sample returning the questionnaire. One-fifth of the zero WTP amounts could be classified as legitimate (Table 5.1). The remaining 798 zero bids (32.9% of the questionnaires returned) were classified as protest responses, in accordance with the criteria reported in Table 5.1. More than half of the protest bidders indicated that they had used forests for recreation, but could not say what they would have been willing to pay for admission. One-quarter stated that they were 'against payment for access to forest recreation' or considered 'access to forest recreation an inherited right'. 'I already pay taxes' was indicated by 10% of the protest bidders as a reason for stating a zero WTP amount. Thus, the reminder in the questionnaire that 'There are no plans to charge the public for admission to recreational areas in Denmark' either went unnoticed or was not accepted as trustworthy by a significant share of the respondents.

In CV surveys using the OE format, a high level of item non-response and protest bidding seems to be common – most probably because respondents are

Table 5.1. Reasons given for stating a zero bid.

	n	%
Valid reasons		
I have not used forests for recreation	72	36
Could not afford to pay	80	40
Other valid reasons	49	24
Total	201	100
Not valid reasons		
Cannot say what I would pay	428	54
Against payment, access is inherited right*	187	23
I already pay taxes*	79	10
Other not valid reasons*	67	8
'Something else' (without specification)	37	5
Total	798	100

*Reason indicated by respondent.

not aided by a payment card or 'take-it-or-leave-it' offer (Mitchell and Carson, 1989; Bateman, 1994). It is not unusual for the rate of item non-response and protest bids combined to be 20–30% for OE WTP questions (Mitchell and Carson, 1989). In the present survey, item non-response and protest bidding together are close to 40% of the returned questionnaires. The use of a self-fill questionnaire may explain why the present survey had a somewhat higher drop-out rate than the 'norm' indicated above. In contrast to face-to-face interviews, the self-fill approach implies that the reasons suggested for not stating a positive WTP amount cannot be concealed from respondents, when asking the WTP question. Hence, there may have been a certain amount of feedback from the follow-up questions, which in turn may have reduced the rate of positive bids.

Overbidding It is critical also to remove implausibly high bids, since sample mean WTP is sensitive to even a single extremely high amount. In the present survey, outliers at the upper end of the WTP distribution were deleted on an *ad hoc* basis, removing all bids exceeding 2000 Danish crowns (DKK).[6] Only four such amounts were recorded.

Other outliers A total of 12 respondents, reporting more than 365 annual forest visits, were classified as outliers and removed from the sample. Finally, in 36 questionnaires the gender indicated by the respondent did not coincide with the gender of the individual the questionnaire had been addressed to. Assuming that these questionnaires had not been answered by the selected individual, this group was also removed from the sample.

The effective sample After the trimming, the reduced sample included 1420 respondents – equivalent to 49.1% of the individuals on the mailing list, or 58.6% of the questionnaires returned.

Representativeness of the effective sample
Based on the trimmed sample, the study attempts to infer a mean WTP for access to forest by the general public. This implies that protest bidders' and other outliers' 'real' WTP is set equal to the mean WTP of the trimmed sample. It is critical, of course, that the reduced sample does not deviate significantly from the general population with respect to determinants affecting preferences for access to forest recreation. Statistical analyses showed that the reduced sample does not differ significantly from the general population with respect to the potentially most important determinants of WTP: region of residence, gender, age, occupation and income (Dubgaard, 1997). Therefore, it seems safe to assume that the trimmed sample is in fact representative of the general public's preference structure.

Willingness-to-pay results
With protest zeros and outliers removed the overall mean WTP was 128 DKK (median = 100 DKK) for a pass giving unlimited access to all Danish forests during a 1-year period. Several studies suggest that OE questions elicit lower bound estimates of WTP (Walsh *et al.*, 1992; Bateman, 1994; Dubgaard, 1996a). In other words, the present result can be regarded as a conservative estimate of WTP for access to Danish forests.

Validity Assessments

The basic difficulty with the CV method is the uncertainties regarding the validity of individuals' stated WTP for a non-market good in a hypothetical transactions setting. Major concerns are whether respondents will be capable of correctly expressing their preferences and whether respondents may engage in strategic behaviour, deliberately misstating their WTP in order to affect policy decisions (Arrow *et al.*, 1993).

Mitchell and Carson (1989) propose a number of validity criteria against which the reliability of the CV results can be examined. Theoretical validity examines whether the CV results conform to economic theory as such, for example if there is a positive relationship between WTP and income, availability of the good being valued, and so on. In the present study, the socio-economic and other behavioural determinants of WTP were examined, using regression analysis. In these experiments, (only) two explanatory variables were found to have significant influence on the WTP pattern: visit frequency (that is, annual visits to all Danish forests) and personal income. The visit-frequency variable was incorporated to test if the CV results comply with the important theoretical validity criterion: that satisfaction from the use of a good (normally) increases with the quantity consumed, but at a diminishing rate. That is, one would expect a positive, but less than unity, numerical value of the elasticity expressing changes in WTP relative to changes in visit frequency. The estimation results supported this assumption, indicating a visit-frequency elasticity equal to 0.17. There was also a positive and significant statistical link between (personal) income and WTP. The estimated income elasticity was 0.28. In a CV study of WTP for access to additional woodland, Bateman *et al.* (1995) found a similar visit-frequency elasticity (0.14) and somewhat higher income elasticities (around 0.65).

AD HOC BENEFIT–COST APPRAISAL OF FOREST RECREATION

Mean WTP of the trimmed sample was used as the relevant welfare measure, and aggregate benefits from access to forest recreation were calculated by multiplying mean WTP for an annual pass with the entire Danish population in the age cohorts included in the survey (that is, people 15–76 years of age). The Danish population in these age-groups consists of about 4 million people. Multiplying this number by mean WTP for an annual pass to Danish forests (128 DKK) yields an estimate of aggregate benefits of about 500 million DKK per year. If children under 15 years (0.9 million) and people over 76 (0.3 million) are 'admitted' at half price, the estimate of aggregate annual benefits increases to about 600 million DKK. As indicated above, these amounts can be regarded as lower-bound estimates of aggregate WTP, because they are derived from OE CV questions.

Tentatively estimated, the social opportunity costs of providing the present level of recreational services equal 100 million DKK per year (Danish Forestry Service, 1995, personal communication). This yields a simple benefit–cost ratio of 5–6 and net social benefits from access to forest recreation equal to around 0.5 billion (500 million) DKK on an annual basis. To illustrate the order of magnitude, gross factor income in the Danish forestry sector is in the order of 1.1 billion DKK. With about 0.5 million ha of forest in Denmark, social net benefits from recreation amount to about 1000 DKK ha^{-1} annually.

COMPARISON WITH OTHER FOREST RECREATION STUDIES

The economic value of forest recreation has been investigated in a large number of foreign studies, especially in the USA, the UK and the other Scandinavian countries (see, for example, Navrud, 1992; Walsh *et al.*, 1992; Dubgaard *et al.*,

1994). However, with one exception, the author has not come across valuation studies which are directly comparable with the present survey in scope, that is, which attempt to measure the general public's WTP for access to the country's entire forest area. Still, using different breakdowns of the benefit estimates, it is possible to make some crude comparisons with the findings of other valuation studies.

The Mols Bjerge Survey

In a recent CV survey, Dubgaard (1994, 1996a) estimated WTP for access to on-site recreation in a silvopastoral landscape in East Jutland, the Mols Bjerge. The OE as well as the DC format was applied. The OE format yielded a recreational benefit estimate equal to about 2100 DKK ha^{-1}. On average, the visitation intensity per hectare in Danish forests as a whole is about half the level found in Mols Bjerge. Therefore, the benefit estimate of about 1000 DKK ha^{-1} for the entire forest area corresponds well with the findings of the Mols Bjerge study.

Willingness to Pay for Non-timber Benefits from Forests in Västerbotten

In format and scope, the present survey resembles a study of forest valuation by Mattsson and Li (1993). Using the CV method, Mattsson and Li estimated WTP for non-timber benefits from the entire forest area in the county of Västerbotten in northern Sweden. The OE format resulted in an annual WTP estimate equal to 2234 SEK per individual, for on-site as well as off-site benefits from the county's forests. On-site benefits alone accounted for about 1500 SEK. This is more than ten times the estimated WTP for access to on-site recreational use of Danish forests. Indeed, there are great differences between Denmark and Västerbotten regarding the size of the forest area and population density. Västerbotten has 3.4 million ha of forest land and only 180,000 inhabitants. The 5.2 million people in Denmark share about 0.5 million ha of forest. Also, consumptive use of forests, in the form of berry or mushroom picking and so on, undoubtedly plays a much greater role in Västerbotten than it does in Denmark. Still, the difference in estimated WTP in the two surveys is remarkable.

Recreational Benefits from Forestry Commission Estate in Britain

In general, the results of the present survey seem to be more in line with the findings of forest recreation studies in the UK. The fact that both countries are densely populated and sparsely forested may explain why this is so.

Using the travel-cost method, Willis (1991) estimated the recreational value of the Forestry Commission (FC) estate in Great Britain. For the FC area as a whole, the study estimated an average consumer surplus of £2 in 1988 prices (20 DKK) per visitor. This figure is not directly comparable with the estimate of WTP for access to forest recreation in Denmark. In most of the FC forests, recreational use was dominated by single visits within a 1-year period (Willis, 1991). In the Danish study, the visitation pattern was dominated by repeat visits. However, at the aggregate level, some (crude) comparisons can be made. For the FC estate as a whole, the study found that the 28 million visitors (per year) gained a £53.0 million consumer surplus from forest recreation. This can be judged in relation to a £71.05 million income from the sale of timber in 1988 and a net expenditure on (non-commercial) forestry recreation and amenity of £8.5 million. Thus, the ratio between income from timber sales and recreational value is about 1.4, and the ratio between recreation value and expenditure on recreation and amenity equals 6.2. For the Danish forest area, these ratios are of the same order of magnitude (1.2 and 6, respectively). Measured on an acreage basis, consumer surplus deviated from £449 (4500 DKK) to less than £1 ha^{-1} in the different FC estate areas. With 1000 DKK ha^{-1}, the Danish estimate of average recreational benefits is within this range.

Marginal Recreational Benefits from New Woodlands in England

In a hedonic price study, Powe *et al.* (1995) investigated the marginal annual benefits from increasing the woodland area in and around Southampton and the New Forest in England. The estimated benefits ranged from £88 to £250 ha^{-1} year^{-1} (900–2500 DKK), depending on the proximity to urban populations. Of course, an estimate of the benefits from an increase in forest recreation opportunities in the south of England is not directly comparable with average WTP for access to existing forests in Denmark. Still, it is possible to illustrate the proportions. With estimated average recreational benefits at around 1000 DKK ha^{-1}, the results from the Danish study are pretty much in line with the findings by Powe *et al.* (1995).

CONCLUSIONS

The present CV survey was the first attempt in Denmark to estimate the recreational benefits from the entire forest area. It is also one of the first large-scale CV surveys of WTP for access to outdoor recreation in Denmark. Recalling that Danish legislation guarantees the right of common access to woodlands and other recreational areas, the results of this study give an indication regarding the applicability of the CV technique in a Danish social and cultural setting. The overall results of the survey are satisfactory. The OE WTP elicitation method, in combination with a self-fill questionnaire, imposes high requirements on respondents. Taking this into consideration, the WTP response rate was satisfactory, with close to 60% of the respondents (returning the questionnaire) giving usable WTP amounts. Also, the WTP pattern recorded was compatible with the most important economic validity criteria, and cross-study comparisons indicate that the estimated mean WTP is within the range of values found in similar foreign investigations.

Totalling 0.5–0.6 billion DKK on an annual basis, the aggregate benefits from forest recreation in Denmark are substantial. However, the fact that net social benefits from access to forest recreation are both positive and significant does not prove that the present level of provision is also socially efficient. Optimal supply of recreational services occurs where the marginal aggregate benefit curve and the marginal aggregate cost curve intersect. Estimating these marginal relationships was beyond the scope of the present survey. Still, the results call for a strong emphasis on recreation in the Danish afforestation policy. Further research is needed to identify the marginal relationships relevant for optimizing the Danish afforestation policy.

NOTES

1 The findings concerning the recreational use of Danish forests and preferences for forest characteristics are reported in Jensen and Koch (1997). More than 90% of the population visit a forest at least once a year and the average (self-reported) frequency is about 38 annual visits per individual (median visit frequency is equal to ten). For the whole population, the number of forest visits was estimated at 155 million annually. The preference investigations indicate that most Danes prefer the characteristics of a 'traditional', extensively managed, deciduous forest.

2 The NOAA Panel was convened by the US NOAA to examine the reliability of the CV method for environmental damage assessments. The panel's report (Arrow *et al.*, 1993) has acquired a pivotal role, defining what are now the generally accepted standards for CV analyses.

3 The WTP measure implicitly assumes that the agents involved do not have previous rights to the resource being valued. If the agents do have such rights (from a philosophical or legal standpoint), the appropriate measure of net economic value would be the willingness to accept (WTA) compensation measure, rather than the WTP measure. However, when both are measured empirically, WTA tends to greatly exceed WTP (for a review of empirical evidence, see Freeman, 1993). The NOAA Panel recommended that the WTP measure be applied to obtain appropriate or conservative CV estimates.

4 Not all commentators agree that protest bids should be removed. For a discussion and further references, see Bateman (1994).

5 Up to three reminders were sent to those who failed to reply. Jensen and Koch (1997) give an overview of the follow-up techniques applied in the form of appeals to respondents who had failed to return the questionnaire.

6 1 DKK = 7.3 European currency units (ECU).

Valuable Landscapes and Reliable Estimates

6

Knut Per Hasund[1]

INTRODUCTION

Swedish agriculture produces private goods for the market and public goods that have been provided free as positive externalities. These public goods are mainly food security (that is, having a domestic production capacity in case of war or trade blockade) and landscape objects or services. How large the landscape values are depends on the size of the cultivated landscape (acreage) and the quality of the landscape, that is, its content of stone walls, field islets, larks and the scent from yellow bedstraw, and, as always in the economist's perspective, on the consumer preferences ascribing these objects values.

This chapter presents an empirical study concerning evaluation of a permanent policy change for preserving ponds, ditches, stone walls, solitary trees, field islets and similar landscape elements of arable land in Sweden. The landscape elements may be considered as non-rival and non-excludable goods that are continuously providing a flow of services. The non-exclusiveness of the services provided by the landscape elements is partly due to the fact that existence values constitute a large fraction of the services.

The items evaluated have the character of a complex good of multiple values and heterogeneous time variability of its components. It is a concrete good: the stone walls and other landscape elements are to a varying extent familiar to and experienced by all residents, if not fully for their biological importance at least in their physical form. There are irreversibility aspects in removing them.

The landscape elements have been declining in number, owing to changes in technology and relative prices, political measures and attitudes among

(eds S. Dabbert, A. Dubgaard, L. Slangen and M. Whitby)

farmers. In the presence of non-rivalry and non-excludability, the market is likely to give a suboptimal provision of these goods. It is thus justified to apply policy measures for preserving the landscape elements, according to the Kaldor–Hicks criterion[2] for welfare maximization, except in two situations. Neither of these was prevalent in Sweden at the time of the study. Policy measures would not be motivated if the elements were provided free as natural endowments or joint products to an extent where social marginal benefits were equal to or smaller than social marginal costs. The second case is a theoretical possibility that would occur if the landowners' (private) optimal production of landscape were to equal the social optimum. In the latter case, the marginal willingness to pay (WTP) of all other citizens is zero or the marginal cost is infinite, at the quantity provided.

Politicians attempting to design a policy that leads toward improved social welfare would require information on the magnitude of the benefits generated by a public-good resource. The only method currently available that can cover use as well as non-use values is the contingent valuation method (CVM).[3] If the WTP or the CVM estimates of WTP fluctuate widely over time, the consequences for the applicability of this approach may be profound. This would especially apply to programmes concerning irreversible changes or sluggishly changing processes, as is very much the case for the preservation of the landscape elements of agricultural land in Sweden. It may, for example, take several hundred years for a certain flora to evolve after constructing a stone wall.

This chapter begins by describing the environmental problem in terms of the depletion of landscape elements and the values that are attributed to them. The CVM surveys are then presented, followed by the estimates of WTP. An attempt to disaggregate the values into various types of elements, reasons for valuing the elements and use or non-use values is also presented. In the final part of the chapter, the temporal stability of the estimates and the temporal reliability of the CVM are analysed and discussed. The temporal reliability is tested systematically by a predictive model and by assaying explanatory variables and coefficients with respect to WTP over time.

THE LOSS OF LANDSCAPE VALUES

Landscapes have always changed. During the last 40 years, however, a drastic change has occurred in the Swedish agricultural landscape which is not comparable to any other time in history. Farmers culvert ditches and remove stone walls and field islets to transform a mosaic of small enclosures into a few, large and easily cultivated fields. In some districts these changes have come to an end, as there are scarcely any more landscape elements to remove, but in other regions the process has just started. This rationalization has previously been actively supported by policy measures. The main motive for the rationalization of the field layout is to decrease the costs of crop production in order to increase farm profits. A policy objective was also to lower the consumer prices.

The field rationalization affects biological diversity, historical remains of cultivation, visual qualities and recreational access of the agricultural landscapes. For example, merely the filling up of ponds and other small wetlands in the fields is threatening 46 species of vascular plants, 30 mosses, 41 vertebrates (animals, birds) and 195 invertebrates, of which 18, 6, 19 and 75, respectively, are endangered (Bernes, 1994; Aronsson *et al.*, 1995). The transformation of the agricultural landscape was considered the most serious threat to biological diversity in Sweden by the Swedish Environmental Protection Board, where the removal of stone walls and other elements constitutes a major factor.

IDENTICAL SURVEYS IN 1989, 1991 AND 1994[4]

The study applied a split-sample approach. Three mail surveys with identical introduction letters and questionnaires were sent out at different times, in 1989, 1991 and 1994, respectively. The three samples were randomly taken from the population of people resident in Sweden aged 16–74. Owing to deceased individuals and non-respondents, the final sample sizes were 337, 195 and 297 potential respondents, respectively.

The questionnaire started with two pages of information about the changes to the landscape, why it is rationalized and some comprehension of its effects. Preceding the WTP question were questions on behaviour and attitudes, in order to make the respondents more familiar with the good and to obtain information on the population. To encourage the respondents to consider their budget constraints, they were asked to allocate the state budget for the environment among other public services and private consumption. An 'open-bid' question was used to elicit the WTP. It asked for the respondents' WTP, and thus not for the WTP of the households. At the end of the questionnaire was a question for respondents stating a zero WTP, where they could mark among seven alternatives for motives of stating zero WTP or express it in their own wording.

The response rates of the questionnaires in 1989, 1991 and 1994 were 59%, 51% and 51%, respectively. A telephone survey of every tenth non-respondent, selected randomly, showed no significant difference in WTP between mail respondents and non-respondent by mail.

THE VALUES OF LANDSCAPE ELEMENTS

Willingness to Pay Estimates

The mean WTP estimates are in the range 600–1200 Swedish crowns (SEK)[5] per year. Corresponding mean standard deviations are 62–224 SEK per year (Table 6.1). If protest bidders and outliers are excluded,[6] the estimated means of WTP in 1989, 1991 and 1994 for preserving all landscape elements of cultivated land in Sweden are respectively 864, 772 and 626 SEK year^{-1} in 1994 values.

The WTP distribution is skewed to the right, with focal points for WTP at 0, 100, 500, 1000 and 2000 SEK. The high standard deviations reflect that there are big differences within the population in the evaluation of landscape elements, with many respondents at zero WTP and many at fairly high bids.

Aggregating the WTP estimates per inhabitant over the population of age 16–74 years reveals that the total WTP for preserving all landscape elements is 4.7 billion (4700 million) SEK year^{-1}.[7] The figure may be compared to the total resources allocated for the preservation.

To get a better idea of the magnitude of the values, the WTP estimates could be distributed per hectare. The WTP for preserving all landscape elements amounts to 1700 SEK ha^{-1},[8] calculated by dividing total WTP by total acreage of cultivated land in Sweden. It should be stressed that this is an average figure and that the frequency of elements, as well as the value of the objects, may vary widely from site to site.

Table 6.1. Mean WTP and mean standard deviation (SD) for landscape elements on cultivated land in Sweden in 1989, 1991 and 1994 (SEK per year in 1994 values).

	Mean WTP*	Mean SD	No. respondents
1989			
Including outliers, including protest bidders	849	126	186
Including outliers, excluding protest bidders	1144	162	138
Excluding outliers, excluding protest-bidders	864	143	112
1991			
Including outliers, including protest bidders	932	185	88
Including outliers, excluding protest bidders	1171	224	70
Excluding outliers, excluding protest bidders	772	83	63
1994			
Including outliers, including protest bidders	669	100	134
Including outliers, excluding protest bidders	801	116	112
Excluding outliers, excluding protest bidders	626	62	103

*Mean WTP *per capita*, per person of age 16–74.

Disaggregated Values

The landscape is a complex good and the preservation of landscape elements concerns multiple values. For policy purposes, it may be interesting to see what constitutes WTP. One dimension refers to reasons for valuing: for example, to what extent is WTP attributable to concern for wildlife or aesthetic qualities? Another dimension of subcategories of values refers to how various physical landscape elements are valued.

To investigate these issues, the survey instrument included two questions subsequent to the WTP question, where the respondents were asked to allocate value points. The respondents were asked to allocate 100 points among various elements, according to what they thought was 'most valuable to preserve'. The results are presented in Tables 6.2 and 6.3.

The alternative 'landscape elements that are of greatest importance to birds and wildlife' was given the most points, on average 28 out of 100. Biological reasons in total account for 50% of the preservation values, while cultural-historic reasons account for one-fifth. Disaggregating the stated WTP for preservation according to the allocation of points, about 200 SEK per capita and year is attributable to landscape elements with cultural-historic values, which corresponds to 450 SEK ha^{-1} of all cultivated land in Sweden.

The relative values of the alternatives conform with the results of Drake (1987), who asked for the respondents' motives for preserving the agricultural landscapes of Sweden from afforestation. The valuation procedure of that study was quite different, so no direct comparison is possible.

Table 6.2. Elements that are most valuable to preserve in terms of their importance for cultural-historical, biological, recreational and other reasons. Mean values for the samples of survey 3, 6 and 9 in 1989, 1991 and 1994, respectively. WTP in 1994 values. Including outliers, excluding protest bidders.

Elements of specific importance for	Value points allocated	Disaggregated WTP* (SEK $year^{-1}$)	Disaggregated WTP (SEK ha^{-1} $year^{-1}$)
Cultural-historical	19.6	200	450
Birds and wildlife	28.1	300	670
Flora	21.7	250	550
Beautiful appearance	18.6	180	400
Facilitate the accessibility	11.3	110	250
Other (respondent specified)	0.8	6	10
Total†	100.0	1000	2300

*Mean of the respondents' disaggregated WTP (D-WTP), which is calculated individually: $D\text{-}WTP_i = S\text{-}WTP_i \times P_{ij}/100$, where $S\text{-}WTP_i$ is stated WTP in valuation question and P_{ij} is respondent *i*'s allocation of value points for alternative *j*.
†May not add up exactly due to rounding.

Table 6.3. Kinds of elements that are most valuable to preserve. Mean values for the samples of survey 3, 6 and 9 in 1989, 1991 and 1994 respectively.

Elements	Value points allocated
100 m stone wall	18.5
100 m ditches	11.8
Five small field islets	17.7
Eight solitary trees	12.1
100 m wooden fences	12.7
100 m headland with grass and trees	15.5
100 m headland, grass and barbed wire	3.0
100 m tractor roads	8.9
Total*	100.0

*May not add up exactly due to rounding.

The respondents were then confronted with the following question:

> Similar to question 10, it may be that there is not support for preserving all the landscape elements. In order to be able to decide which elements should be preserved in the first place, it is important to get to know how you value the different kinds of landscape elements. As in the preceding question, You have 100 points to distribute freely among the alternatives that You find most important to preserve.

Stone walls and field islets are the highest-valued elements, but ditches, solitary trees, wooden fences and headlands with grass, bushes and trees are also highly valued. Tractor roads and especially headlands with merely grass and barbed wire are not considered as important to preserve.

Only 1% of the respondents allocated the same amount of points (12.5) to each of the eight alternatives. Still, 16% allocated the value points in similar proportions, by giving each of the eight alternatives at least 10 points. A minority of 18% had favourite elements, allocating at least 50 points to a single element.

It has been demonstrated in several studies with more than one good to evaluate, that the order of the alternatives influences the WTP estimates (see, for instance, Frykblom, 1997). With the intention of controlling the order-effect bias, the order of the alternatives in questions 10 and 11 was shifted across three subsamples. The analysis clearly shows that there was no order effect with this question design.

Discussion on the disaggregation of values

These allocation questions should provide information on the relative values of the elements, but the outcome should, of course, be interpreted with care. The weight given to a specific alternative is in general negatively correlated with the total number of alternatives. For example, if avenues were added to the alternatives, it is likely that stone walls would be given fewer value points.

Secondly, the supply of ready-worded alternatives tends to attract interest to the phenomena they represent. This is irrespective of the existence of the open alternative 'other', where the respondents have the opportunity to state the very same kind of object. It is thus likely that 'elements that are good wind shelters' would have received more value points if it had been added as a ready-worded alternative. Thirdly, presenting separate alternatives would in general give more points in total to a category than a single, wider alternative would. The combination of the two alternatives 'birds and wildlife' and 'flora' is likely to result in more points given to biologically valuable elements than if there had just been one alternative for 'elements that are of large importance for birds, wildlife or the flora'. As usual, the results should be understood in relation to their origin (Mitchell and Carson, 1989).

The quest for more detailed information on what constitutes the values of the landscape elements in monetary or at least quantitative terms must not entail misleading procedures. Having more than one WTP question in a survey is not to be recommended, since it may give rise to sequencing bias (Diamond and Hausman, 1994). This may in turn imply problems of confusing the respondents' comprehension of scenarios, budget constraints, inclusiveness and so on, and influences by emotional or cognitive links across questions (Mitchell and Carson, 1989; Diamond and Hausman, 1994). Hanemann (1994a) agrees that sequencing effects exist, but writes that they are explained by substitution effects and diminishing marginal rates of substitution.

The sequencing problems could be avoided by split-sample surveys, but another embedding problem may still persist, the scope or part–whole inconsistency. Kahneman and Knetsch (1992) claim that: 'willingness to pay for the same good can vary over a wide range depending on whether the good is assessed on its own or embedded as part of a more inclusive package'. Another study, by Kemp and Maxwell (1993), which disaggregated WTP for a broad group of government programmes by dividing and subdividing, obtained the estimate $0.29 for minimizing the risk of oil spills in Alaska, as compared with $85 for the sample that valued the same programme directly. These studies have been criticized on survey-procedural and theoretical grounds (Smith, 1992a; Hanemann, 1994a). The existence of scope inconsistencies has also been repudiated by Hanemann (1994a) on empirical grounds, with reference to a large number of studies that elicited proper scope patterns of WTP. In any case, the disaggregated WTP estimates of Table 6.3 must not be confused with WTP estimates for the categories of elements.

The procedure of letting the respondents allocate value points avoids the problems of sequencing WTP questions. Another advantage is a high response rate, higher than the preceding WTP elicitation question. The many odd distributions of points may indicate that the respondents have also been taking the allocation task quite seriously. It is as yet untested whether or not the validity of the responses for allocating value points is higher than that of answering WTP questions.

Allocating value points should give cardinal estimates of relative values. In case these are a complete set of additive subvalues to the WTP for preserving all landscape elements, the WTP for the allocation alternatives could be derived (as in Table 6.2). The possible existence of scope inconsistency suggests that these derived WTP estimates are taken for what they are. The experience from Kahneman and Knetsch (1992) and some other studies indicate that the WTP for the stone walls or any of the other elements would have been higher if the survey addressed only this matter, instead of asking for WTP for preserving all elements and then multiplying by the fraction of allocation points. As noted previously, other studies suggest that there is no such difference.

Scope effects could be avoided by another survey procedure, where a WTP question for one of the alternatives is combined with a value-point allocation question for the complete set of alternatives. Substitution effects are thus avoided, but direct value transferability would still require that there were no budget effects.

If the subcategories to the valued item are complementary goods, no allocation of value points and no value transfers are possible. When the conditions of additivity and a complete set are satisfied, disaggregating WTP by value points could be a promising approach for providing policy-relevant information.

Use and non-use values

Another dimension of disaggregation that may also be relevant for policy design is what constitutes the WTP in terms of use or non-use values. If, for example, the direct use values dominated, it could be argued that relatively more resources should be allocated to frequently visited or viewed areas.

To obtain an idea of the relative importance of use and non-use values, the enquiry form had the double question:

> On question 7 You stated the amount that You are willing to pay for preserving the landscape elements. About how much of that amount would You have been willing to pay if it only concerned areas that You yourself visit or stay in, and how much was assigned because you value the fact that those landscape elements also exist in the rest of the country?
>
> 9.A Answer: About ________ Swedish crowns of the amount that I stated in question 7.A was to protect the landscape elements in my home district and where I usually spend my vacation.
>
> 9.B Answer: About ________ Swedish crowns of the amount that I stated in question 7.A was to protect the landscape elements in the rest of Sweden.

The wording of the question does not comply with an exact or consistent demarcation between the respondents' use and non-use values, but the answers to 9.A and 9.B should give an approximate estimate of them, respectively. The respondents may of course ascribe non-use or existence values also to the landscapes that they enjoy. On the other hand, they may also be occasional, direct consumers of landscapes in other parts of Sweden. The reason for wording the question this way was that it would be inappropriate to ask directly about use and non-use values.

With that in mind, it can be noted that the mean WTP for the 'own landscapes' of question 9.A (≈ use values) for 1989, 1991 and 1994 was 550 SEK (in 1994 values). The mean disaggregated WTP for 'landscape elements in the rest of Sweden', as in question 9.B (≈ non-use values), was 680 SEK. It corresponds to 45% and 55%, respectively, of the WTP for preserving the landscape elements. About 5% of the respondents stated that the full WTP was for 'non-use' landscape elements, but less than 2% stated that the full WTP was for their 'own landscape elements'. All the results were stable over the years.

In conclusion, the use and non-use values appear to be of about the same magnitude for preserving the elements of the agricultural landscapes, maybe with a little preponderance for higher non-use values.

TEMPORAL RELIABILITY OF CONTINGENT VALUATION ESTIMATES

The CVM surveys of WTP for public goods are mostly performed in order to get a quantitative economic basis for court proceedings or policy decisions. In either case, credibility, validity and reliability of the estimates are essential but controversial. One specific issue addressed is the temporal reliability of the CVM estimates. The pivotal National Oceanic and Atmospheric Administration (NOAA) Panel is concerned regarding this in their *Guidelines for Value Elicitation Surveys*. Accordingly, they recommend an 'adequate time lapse from the accident' and that 'time dependent noise should be reduced by averaging across independently drawn samples taken at different points in time' (Arrow *et al.*, 1993). Based on the time-consistency issue, some researchers have questioned the CVM in general, arguing, for example, that it is 'typically the result of snap judgements based upon superficial emotional appeals broadcast on television' (Magleby, 1984). This assessment contributed to Magleby's proposition that referendums 'can be a most inaccurate barometer of their opinions' (Magleby, 1984, p. 144).

Reliability refers to the reproducibility, or the consistency, of a result. For CVM estimates, it is thus inversely related to the extent to which the variance is caused by random sources or non-identified factors. Temporal reliability of a CVM survey implies that it should reflect stable values when preferences, incomes and explanatory variables have not changed, and should reflect changes in values when these factors have changed.

Previous Research on Intertemporal Consistency

Two techniques have been used for testing the temporal reliability: the test–retest, which repeats the survey for the same respondents, or the split-sample test, which uses distinct samples from the same population for each survey.

The hypothesis that the estimates of WTP are reliable over time has not been rejected in any of the available references. In all studies but one, WTP estimates appeared stable over time. This may support the policy usefulness of values hence estimated, by not finding evidence for the possible cyclic, the irregular, or the shock–subside temporal patterns of WTP.

The reliability is general for all qualities tested so far. It has so far been verified for public and private goods, for the dichotomous choice as well as the open-ended question formats, for environmental goods of mainly use or mainly existence values and for familiar and unfamiliar goods.

Still, generality outside these cases is not granted, and doubts may be cast about the results due to major or minor flaws in the research. Some studies (for example, Loomis, 1989; Kealy *et al.* 1990) suffer from a possible bias toward stability in estimates by using the test–retest technique, which may give recall effects. The respondents are then preconditioned by their previous responses. Another potential flaw is the use of ordinal or subject-unique scales for inter-respondent comparisons and regression estimation. Loomis (1990), Whitehead and Hoban (1996) and Kealy *et al.* (1990) use such measures for attitude or preference variables. Erratic results may also originate from survey design: by giving incentives for free-riding (Kealy *et al.*, 1990) or strategic behaviour (Loomis, 1990), by field-price censoring (that is, presenting an experiment commodity price higher than the market price) (Kealy *et al.*, 1990), by not allowing for negative WTP (Reiling *et al.*, 1990), by the valuation question inviting hypothetical bias (Bergland *et al.*, 1996), by posing multiple valuation questions to the respondent (Loomis, 1990; Carson *et al.*, 1992), and by inducing possible anchoring effects (Teisl *et al.*, 1995). Further flaws include not having identical surveys (Carson *et al.*, 1992), having low response rates (Loomis, 1989) or using only students as evaluating subjects (Kealy *et al.*, 1990). As usual, the results of dichotomous-choice studies, such as Teisl *et al.* (1995), are not independent of which parametric assumptions are made. These reservations do not necessarily imply that the final conclusions of temporal reliability are invalid, especially since the evidence in many studies is strong.

Methodology for Testing Temporal Reliability

The questions addressed in this chapter concern consistency over time spans of a few years. Is real[9] WTP stable or does it fluctuate from year to year? Does estimated WTP correctly reflect changes in real WTP, that is, does it increase as real WTP increases, and remain stable or decrease if real WTP does so?

Combining these two questions gives four possible outcomes depending on whether they are answered by yes or no (Table 6.4).

Temporal consistency would correspond to either quadrant 1 or quadrant 4 in Table 6.4. Since real WTP cannot be directly observed by the CVM, a test of the consistency has to be performed versus real behaviour, that is, in experiments or in parallel markets. Such an experiment was conducted by Kealy *et al.* (1990) in a 2-week test–retest, giving support for the temporal reliability of CVM in that setting. The applicability of this approach is limited, since most surveys attempt to value public goods. Another approach that could give evidence for or against the reliability is to test the method's internal consistency and consistency versus theory.

The basis for this approach is to compare if estimated mean WTP and its determinants have changed over time and according to what economic theory would predict. Using the linear-regression model as an example for illustrating the principal problem, the mean WTP could be calculated by the equation:

$$\overline{WTP_t} = \beta_0 + \beta_{1t}\overline{I_t} + \beta_{2t}\overline{P_t} + \ldots + \beta_{nt}\overline{S_t} + \overline{\varepsilon_t} \quad (1)$$

where $\overline{WTP_t}$ is mean WTP in year t (t = 1989, 1991 or 1994 in this study), $\overline{I_t}$ is mean income in year t, β_{1t} is a coefficient expressing the impact of income on mean WTP, $\overline{P}$ is the preference or taste variable, $\overline{S}$ is a variable in the set of sociodemographic variables determining the WTP. Mean $\overline{\varepsilon}$ is the divided sum of all respondents' error terms, covering the fraction of WTP that is not explained by the included determinant variables. For the OLS method, its expected value is zero.

Temporal reliability would require that either the factors determining WTP have not changed and the estimated WTP has stayed constant, or that the determining factors have changed and the estimated WTP reflects this change. Three cases can thus be distinguished, a stable WTP and stable determinants, a stable WTP and changes of the determinants that offset each other, and a changing WTP that reflects changing determinants.

When mean estimated WTP has not changed, a sufficient but not necessary condition for reliability is if the mean of each determining variable has not changed and each coefficient remains unchanged. In this case, evidence against the reliability of CVM comes by a rejection of the null hypothesis:

$$H_0 : \beta_{i,1} = \beta_{i,2};\ \overline{X_{i,1}} = \overline{X_{i,2}} \mid \overline{WTP_1} = \overline{WTP_2};\ \forall\, i \quad (2)$$

Table 6.4. Possible combination of stable or fluctuating real and estimated willingness to pay over time.

		Real WTP	
		Stable	Fluctuating
Estimated WTP	Stable	1	2
	Fluctuating	3	4

where $\overline{X}_{i,t}$ represents the mean of determining variable X_i in year t.

A general condition[10] for reliability when estimated mean WTP has stayed constant is when changes in the determining factors offset each other so that mean WTP is not expected to alter. For the linear case, it means testing the null hypothesis:

$$H_0 : \Sigma\beta_{i,1}\overline{X}_{i,1} = \Sigma\beta_{i,2}\overline{X}_{i,2} \mid \overline{WTP_1} = \overline{WTP_2} \quad (3)$$

In the second case, when the determining factors have changed and the estimated WTP reflects this change, reliability is supported by not rejecting the null hypothesis:

$$H_0 : \Sigma\beta_{i,1}\overline{X}_{i,1} < \Sigma\beta_{i,2}\overline{X}_{i,2}\beta_{i,1} \mid \overline{WTP_1} < \overline{WTP_2} \quad (4a)$$

or

$$H_0 : \Sigma\beta_{i,1}\overline{X}_{i,1} > \Sigma\beta_{i,2}\overline{X}_{i,2}\beta_{i,1} \mid \overline{WTP_1} > \overline{WTP_2} \quad (4b)$$

Thus, analogous to the case of an increasing or decreasing real WTP and a corresponding change in estimated WTP (quadrant 4 in Table 6.4), changing real circumstances should be reflected by an altered CVM-estimated WTP. To be consistent with economic theory, an increase in the preference variable or its coefficient should, *ceteris paribus*, be related to a higher WTP. If there is no reason to expect the good to be inferior, an increase in incomes should analogously result in an elevated WTP.

The procedure presented below first tests whether WTP has been stable over the years or not, and then examines the reliability of the method by testing if the temporal patterns of WTP and its determinants are consistent.

Stable or Declining Willingness to Pay?

Do the presented estimates show that WTP has changed over time, or are the differences within what could be expected from normal sample variation? The hypothesis that WTP has not changed over the years was tested by various methods comparing parameters or distributions. For a full presentation of the tests and their results see Hasund (1997).

Comparing mean willingness to pay across years

The temporal stability of WTP was checked by pairwise t-tests comparing mean WTP for each of the 3 years. The tests indicate that there has been no change at mean WTP over time. The null hypotheses of equal means are not rejected on the 5% significance level, whether outliers and protest bidders are excluded or not.

Another kind of test is based on pooling all data and introducing dummy variables for a year or pair of years. If the coefficient of the dummy is significantly different from zero, this can be interpreted as evidence that the new variable explains some of the measured differences in WTP, that is, there has been a change in WTP between the years. Such tests were conducted using three different samples (excluding outliers and protest bidders, or not) and various regression methods.

The null hypothesis that means WTP is stable over the years cannot be rejected at the 5% significance level. All but one of the coefficients for the year variables are clearly within the 95% confidence intervals. The only test indicating on the 5% level that mean WTP has changed is the OLS for a pooled data set when outliers and protest bidders are excluded, indicating that WTP has declined between 1989 and 1994 by 269 SEK. In the other tests, the result of a time-stable mean WTP is supported whether outliers or protest bidders are excluded or not, and independent of whether OLS or Tobit is used for testing.

Which test would be most appropriate to apply depends on the character of the evaluated good, the purpose of the study and the reliability of the bids defined as outliers or protest bidders. In this study, all arguments are in favour of advocating the Tobit model, with true zero bids separated from negative WTP,[11] which would indicate unambiguously that mean WTP has not changed.

It is noticeable that the coefficients of the year variables for 1991 and 1994 are all negative, which may indicate a non-increasing time trend, though not large enough to be significant.

A third type of method, based on ranks and associated randomization, is the Wilcoxon–Mann–Whitney (WTW) test. The two-tail WTW test is able to detect differences that the *t*-tests do not, since observations with high values will inflate the estimated variance in the denominator of the *t*-statistic. In this study, the WTW tests do not reject the null hypotheses on the 5% significance level. The result is clear, whether outliers and protest bidders are excluded or not. The medians of the samples of 1989 and 1994 do not differ significantly.

Comparing the distributions of willingness to pay across years

Simply comparing the means may not tell everything about the changes over time, since, for instance, it is possible that two (or more) distributions by coincidence have approximately the same mean, while their distributions may differ significantly. Since the standard deviation of the WTP distribution for landscape elements proved to be high, changes over time may also be hidden by the sample variation. Testing for changes in the distribution may thus be stricter than tests on the mean.

To check if the distributions of WTP are the same for the 3 years, the Kolmogorov–Smirnov (KS) and the Cramer–Von Mises (CM) tests were conducted on samples, either including or excluding outliers and protest bidders.

The computed values of the CM test *t* statistic were about 0.3, which is below the critical value for the two-sided test at the 5% significance level. The null hypotheses of equal distributions are thus not rejected, neither for samples including outliers and protest bidders, nor for samples excluding them. This result is a strong indication that WTP has not changed over time. The result is obtained in spite of the methodological problem that is related to ordered ranking tests, due to possible effects of nominal focal points in combination with inflation. Many responses tend to be focused to rounded numbers, such as 100 or 1000 SEK. In the ordered ranking, the inflated values of 1989 of each focal point come after the 1994 values of the same nominal focal point. This tends to increase the differences of cumulative frequency between samples of 2 different years. The test statistic thus becomes unwarrantedly exaggerated, which leads to overrejecting a hypothesis of stability.

On the contrary, the KS tests induce a rejection of the null hypotheses. An interpretation of this contradictory result may be that there has been some dislocation within the distributions, but not large enough to have any significant impact on the mean WTP, given the big deviations in WTP from the mean. A possible explanation is the sensitivity of the KS tests to spikes in combination with inflation.

In summary, there is predominant evidence that mean WTP has not changed significantly across these years.

Testing the Temporal Reliability

Knowledge about CVM's temporal reliability would either require tests of estimated WTP against real behaviour at different times or, when lacking that opportunity, tests of changes in WTP against its determinants. As discussed previously, the latter approach would imply testing one of three alternative hypotheses. Since the tests discussed in the previous section indicate that WTP has been stable over time, the relevant tests could be restricted to:

I. Tests against real WTP.
   A. Market behaviour.
   B. Experiments.
II. Synthetic tests (Construct reliability).
   A. Theoretical consistency.
   B. Empirical consistency.
      1. Testing $H_0 : \beta_{i,1} = \beta_{i,2}; \overline{X}_{i,1} = \overline{X}_{i,2} \mid \overline{WTP_1} = \overline{WTP_2}; \forall\ i$.
         (a) Variables: *t*-tests.
         (b) Coefficients.
            - Simultaneously.
              Chow tests.
              Log-likelihood tests.
            - Individually.
      2. Testing a predictive model $H_0 : \Sigma\beta_{i,1}\overline{X}_{i,1} = \Sigma\beta_{i,2}\overline{X}_{i,2} \mid \overline{WTP_1} = \overline{WTP_2}$.

Theory would predict that, unless the landscape is an inferior good, the income elasticity should each year be positive, which implies $\beta_I > 0$, for the model to be reliable. The coefficient of the preference variable is also expected to be positive.

One empirical approach is to test whether the determinants of WTP have been stable or have changed. The temporal reliability could thus be tested in a three-step procedure, where the first step would be to test if estimated WTP has changed. The second step could be to test whether the income, age, preferences and other determining variables have changed. Finally, it should be tested whether and how the behavioural relationships, the β coefficients, have changed.

When mean WTP has been stable, as is the case here, an indication of reliability would be if the determining variables, take the same values $\overline{X}_{i,t=1} = \overline{X}_{i,X=2}$. The variables that are considered relevant include gender, resident status, age, income and preferences for the environment. The preference variable is derived from a survey question asking the respondents to state how many SEK of the state budget they think the state should allocate annually for environmental protection per capita. This variable has the advantage of being cardinal and suitable for interpersonal comparisons. According to the *t*-tests on sample means, none of the determining variables had changed significantly from 1989 to 1994, thus supporting the hypothesis of reliability.

The third step in this reliability-testing procedure is the testing of the coefficients of the sample regressions, either jointly by a Chow test or log-likelihood tests, or separately by testing each pair of coefficients without any restrictions on the other coefficients. Referring to eqn (2), the null hypothesis for the Chow test would be:

$$H_0 : \beta_{i1} = \beta_{i2} \ \forall i \text{ simultaneously} \quad (5)$$

The test statistic is calculated by regressing the data for each year separately and for the pooled data set, and then comparing the residual sums of squares. It follows an *F* distribution (Greene, 1993, equations 7–14 and 7–20). The Chow tests support the hypothesis that the behavioural relations, the β_i, have not changed when regressing only on variables that were significant in the original regression. When including other variables, such as education and gender, which do add to the total explanatory power of the model but are of little significance by themselves, the Chow tests reject the hypothesis of equal coefficients.

The log-likelihood tests reject the hypotheses that the models of 1989, 1991 and 1994 are structurally identical when regressing the full set of explanatory variables on WTP. On the contrary, regressions on log of income and the other variables with respect to log of WTP give evidence that there has been no change of the coefficients.

Creating new variables by multiplying year dummies with the explanatory variables makes it possible to distinguish which variables have changed. Thus, labelling the variables and examining the regression outputs reveal that it may be the income elasticity, β_I, that has decreased slightly, and the coefficient of the age variable has increased somewhat.

Another, holistic approach is to test $H_0 : \Sigma\beta_{i,1}\overline{X}_{i,1} = \Sigma\beta_{i,2}\overline{X}_{i,2} \mid \overline{WTP_1} = \overline{WTP_2}$. It was done by first running Tobit regressions to estimate the standard deviation of WTP for each year and also the mean values for the variables, $\overline{X}_{i,t}$. These mean variables were inserted for creating a representative average respondent for 1989, 1991 and 1994, respectively. Predicted WTP for the average respondents was then calculated by running new Tobit regressions. Finally, *t*-tests were conducted to see if predicted WTP had changed. The test gives support for CVM being temporally reliable. The null hypothesis is not rejected, which indicates that there has been no change in the determining factors, neither in the coefficients nor in the variables, or, if there have been changes, that these offset each other in such a way that makes the result still consistent.

CONCLUSIONS AND DISCUSSION

This study indicates that the estimated WTP has been stable from 1989 to 1991 and 1994. Compared with previous studies, this is a relatively long test period. There is also evidence that the CVM is temporally reliable. The conclusion is based on tests comparing the temporal patterns of WTP with its determining factors and with the outcome of a prediction model. It is stated that checking the reliability just by tests on the variables or the coefficients is insufficient. The experience from previous studies of a reliable WTP is thus reinforced and widened to include these kinds of environmental goods.

Reflections on Stability and Volatility

What are the arguments suggesting that real or estimated WTP should be stable, and what are the arguments suggesting that they should fluctuate (Table 6.4)?

Besides shifts in determining factors, such as income and income elasticity, that have been tested for, a possible argument for a fluctuating real WTP would be if there were drops in expected future income and prospects of future wealth. Reduced incomes may, for example, shift expectations down and cause pessimism in a way that is not directly visible from the income data. It might be that it would cause an even more pronounced shift down in WTP for some goods, especially for goods perceived as luxury goods or programmes with long-term economic commitments. For risk-averse people, an increasing risk of becoming unemployed and getting a reduced income may cause WTP to drop even more than could be explained by just the reduction in expected income.

Another argument for a fluctuating WTP would be if some respondents are favourable in general toward farming or state support of farmers or have aversions to it. Changes in agricultural policy, in support to farmers or in economic conditions for agriculture for reasons other than landscape preservation may then change people's WTP for a programme that compensates farmers for not removing landscape elements, whether or not they value the elements differently. Schkade and Payne (1993) found that some respondents estimate the cost of providing the good rather than engaging in a self-examination, testing their preferences and paying ability for the good being valued. If the costs of providing the good change, a possible effect could then be that the stated bids would also change. Since the policy and the economic conditions for Swedish agriculture shifted drastically from 1989 to 1991 and from 1989 to 1994, the stability of the WTP estimates for preserving the landscape elements would, however, indicate that people do state their true, maximum WTP for having the landscape elements also in the future, and do not take into account the changing cost situation for preserving them. The stability of the estimates would then indicate that the WTP is not heavily influenced by aversions or favourable attitudes to farming or compensating farming when evaluating an environmental good or its preservation programme. An alternative explanation would be that many people are not informed about the drastically changing agricultural policy or economic conditions of agriculture.

Possible arguments for real WTP being stable may be attributable to the character of this particular environmental good. People may possibly realize that the landscape processes relate to a long-run perspective, with slow speed and some time-lags, and thus that conservation cannot be changed from day to day or year to year with changing short-term conditions. They may consider that when stating their bids and have stable preferences. The values attached to agricultural landscapes are to a large extent existence values and bequest values. An as yet untested hypothesis is that this fact may put them into a more long-term perspective and make them more stable over time. Another hypothesis may be that the cultivated landscape is not a luxury good, as often expected. The landscape is a tangible, visual and familiar good. For many, it is probably also a non-negligible part of their identity, as experienced in its surroundings and historical settings.

The estimated WTP could also be stable because of anchoring effects. It has been suggested that the respondents, when creating their value for the environmental good and stating their bids, use available information for figuring out what would be their appropriate WTP. The existence of figures in the survey instrument may thus have a large anchoring impact on the stated bids (see, for instance, Tversky and Kahneman, 1974; McFadden, 1994; Green *et al.*, 1995; Jacowitz and Kahneman, 1995). A levelling impact on the WTP estimates could not be excluded in this study, since the respondents were given identical questionnaires in 1979, 1991 and 1994. This would not be a typical anchoring effect related to presented figures, but a more general survey instrument-related semianchoring effect, what may be referred to as a grapnel effect. However, a CVM estimate should generally be considered in relation to the survey instrument. Any interpretation of the results has to be stated in the terms: 'Given this survey instrument, the WTP for . . .' With this understanding, the grapnel effect becomes irrelevant and the estimates show, in this study at least, that the WTP has been stable.

The Political Serviceability of the Contingent Valuation Method

The CVM's policy serviceability as concerns temporal reliability involves a scientific issue and a political issue. While the scientific issue mainly concerns methodology and economic theory, the political issue concerns the temporal patterns of WTP and its determinants. To be properly useful for policy decisions, a first prerequisite is that the contingent valuation (CV) estimates are temporally reliable. In addition to that, the temporal patterns of WTP, preferences, determining variables and the relations between these variables and WTP are important.

Because of the slow or irreversible nature of the landscape processes, the policy measures should be stable over time in order to achieve long-run efficiency. Given that reliability is confirmed, stable conditions and a stable WTP would give a more simple basis for decisions on the magnitude of policy measures. On the other hand, if WTP does fluctuate, the task becomes more complicated. Information would be required on the functional form of WTP over time so that net present value could be calculated such that an appropriate magnitude of the policy measures can be set. If the WTP pattern is not linear, information is also needed on the temporal location the study is conducted at. Is the survey conducted at the peak of WTP or at another phase? In conclusion, repeated surveys are required if there are reasons to believe that WTP is not stable over the duration of the policy measures.

Relating the issue to the WTP-determining factors, the consequences for the serviceability of CVM will be different whether it is the preferences, the socio-economic variables or their coefficients that change or not. If education, income or any other socio-economic variable that influences WTP changes, the WTP estimate and policy measures could be easily adjusted to the new situation. It would not be necessary to conduct the surveys again. If, instead, the preferences are not stable, the CVM may still be reliable, but its products will have a limited tenability. Similarly, if it is assumed that the relation between WTP and income or some of the other coefficients might alter, there is a case for repeating the surveys.

In this study, the policy serviceability of CVM estimates is confirmed by the evidence of a temporally reliable estimation method, combined with the indications of a stable WTP, stable preferences and consistent structural relationships.

The experience of this study could increase our confidence in the usefulness of CVM estimates for long-run policy measures, at least for this kind of environmental goods, and at least for time spans in the magnitude of a decade. Whether this conclusion is transferable to other populations or unique to the Swedish population cannot be judged from this study.

NOTES

1 The author would like to thank Professor Per-Olof Johansson at Stockholm School of Economics for valuable help in conducting the study. I would also like to thank Professor Glenn Harrison at the University of South Carolina, Professor Alan Randall at Ohio State University and Professor Olvar Bergland at the Agricultural University of Norway for their contributions. Special thanks also to my colleagues Peter Frykblom, Fredrik Holstein and Dan Vadnjal for their constructive ideas and comments. The study is cofinanced by the Swedish Environmental Protection Agency.

2 The Kaldor–Hicks criterion for welfare maximization states that a measure is justified if the winners are able to compensate the losers, so that everybody is on a higher utility level after the change. The compensations do not actually have to take place; a potential for compensation is sufficient.

3 For a description of the contingent valuation method, see, for example, Mitchell and Carson (1989).

4 For a more complete presentation of the surveys and tests, see Hasund (1997).

5 1 European currency unit (ECU) ≈ 8.5 SEK; $US1 ≈ 7.3 SEK.

6 Protest bidders are defined as respondents having reported protest motives for stating a zero WTP. Outliers are defined as respondents stating a WTP higher than 3% of income.

7 Excluding outliers and protest bidders; SEK in 1994 values.

8 Average for mean estimates of 1989, 1991 and 1994; excluding outliers and protest bidders; SEK per year in 1994 values.

9 The term 'real' refers in this context to situations involving actual monetary transactions, in contrast with hypothetical situations.

10 The previous condition of testing where no coefficient or variable has changed is a special case of this general form.

11 The Tobit regression would be more correct than the OLS, since we have a distribution of censored data with a spike at zero and no reported negative bids. For this particular public good, it is not unrealistic that some respondents actually have a negative WTP for the preservation programme. Other respondents may have a corner solution with WTP at zero; they are indifferent. To obtain correct estimates, these two groups have to be treated differently in the regression. In this study, they were separated by follow-up questions on motives for stating zero WTP.

Cost–Benefit Analysis of Landscape Restoration: a Case-study in Western France

7

François Bonnieux and Philippe Le Goffe

INTRODUCTION

The modernization of agriculture has brought about significant changes in the landscape of western France. Larger machinery requires larger fields, demanding the removal of hedgerows. Landscape has become less attractive for recreation and tourism, and a number of habitats are endangered. Many people deplore this dramatic evolution and ask for policy initiatives directed towards the conservation and restoration of traditional landscapes.

This chapter considers the issue of landscape restoration in Lower Normandy. A project is considered, which would be carried out under the supervision of a natural regional park (NRP), which is in charge of managing an Environmentally Sensitive Area (ESA). This ESA has been primarily designated to protect wetlands, in some of which bird habitats of international interest can be found. It extends over 125,000 ha of low-lying land formed by the flood plains of five rivers and 100,000 ha of dryland. This comprises a number of small plots, planted hedges and ditches, the scenery being characterized by a patchwork of meadows. This traditional *bocage* has been moulded by farmers' activities over the years and is considered to be beautiful and attractive. Elm trees were symbolic of the area and accounted for more than 60% of the total number of tall trees.

The predominant farming activity is dairying, while beef cattle are also raised. The trend towards more intensive dairy farms has led to an increase of silage and maize crops, at the expense of hay and permanent pasture. At the same time, the development of consolidation of holdings has changed the character of the grassland and led to a reduction in the area suitable for

 The Economics of Landscape and Wildlife Conservation (eds S. Dabbert, A. Dubgaard, L. Slangen and M. Whitby)

breeding birds. All such changes, occurring in both dry- and wetlands, radically alter landscape and wildlife value. The dramatic decrease in planted hedges has become greater since the late 1970s, due to Dutch elm disease, which led to the death of elm trees.

A pilot scheme to restore planted hedges was initiated in 1992. Its general objective is to achieve a network of 4-ha plots, with 150 m ha^{-1} of hedgerows. This scheme has been implemented through standard agreements, covering a 5-year period, between individual farmers and the NRP. The NRP offered technical assistance and grant aid (4000 French francs (FF)[1] ha^{-1}) for the planting, regeneration and management of new and existing hedgerows. This scheme has been restricted to a limited area and only 30 farmers have so far been enrolled. The first purpose is to demonstrate the feasibility of replacing elm trees with other tall trees, such as ash, maple, false acacias and wild cherry. The second purpose is to assess the actual cost of replanting and maintaining hedgerows.

THE COSTS AND BENEFITS OF THE RESTORATION PROJECT

The NRP is now faced with the possible generalization of the pilot scheme, over the whole area affected by the disease. The restoration project would extend over 16,000 ha and would consist of restoring 2400 km of hedgerows over a 10-year period. As unitary costs are derived from the pilot scheme, the overall cost of the project is readily assessed. Cost transfer is used and two categories of costs are considered: (i) planting and regenerating costs; and (ii) maintenance costs (Table 7.1). The former are borne during the first 10 years, whereas the latter are spread over the whole life of the project, that is, 100 years.

The issue of benefits is more intricate, because they include private benefits as well as public benefits. First of all, hedges are multiple-use assets, which provide marketable timber and firewood. They also furnish shade, serve as

Table 7.1. Methods to value the costs and benefits of the project.

	Valuation technique
Costs	
Planting and regenerating	Value transfer
Maintaining hedges	Value transfer
Benefits	
Firewood harvest	Market pricing
Timber harvest	Market pricing
Agricultural productivity	Value transfer
Amenities and recreation	Contingent valuation

wind-breaks and secure soil against rapid water runoff and erosion. They provide habitats for birds and other animals. In their many uses, planted hedges yield benefits both when cut and while standing, so benefits flow from extraction and conservation. In addition, they provide a possible contribution to reducing global warming by supplying a source of fuel wood and so reducing the use of fossil fuels (Fankhauser, 1994).

As a commercial crop, trees are viewed in an agricultural setting with relatively long commercial maturation periods (Neher, 1990). In a steady state, each cohort experiences the same life cycle over a rotation period. The annual harvest of each cohort is a repetition of the previous harvest of trees planted 1 year earlier. The number of trees is selected so that a cohort is cut each year and the yield of the planted hedges is sustained. As regards firewood, the rotation period is 15 years and a yearly harvest of 25 m^3 km^{-1} is expected from the fifteenth year until the end of the project. As regards wood for timber, the rotation period is 60 years and a yearly output equal to 60 m^3 km^{-1} is expected over 10 years (that is, from the 60th year until the 69th year). Market prices are used to value both firewood and timber wood (Table 7.1).

Farmers enjoy a number of direct benefits, which are well documented in the literature. Two which are particularly significant are taken into account. First of all, there is a 10% decrease in the rate of occurrence of mastitis, which affects one dairy cow out of four. Second, there is an increase in the yield of permanent pasture, resulting in an increase in forage output and therefore a drop in the feeding costs of dairy cows. The increase in forage output is around 100 kg dry matter ha^{-1}. For maize, there is no significant overall effect on observed yields, with a 2–3-m-wide strip around plots producing a decreased yield, which is compensated for by increased yields elsewhere. Finally, the overall effect on agricultural productivity is determined by transferring values obtained in other studies.

There are several approaches to valuing the public benefits the hedges provide. These benefits cover a variety of environmental goods, some of which are public goods, such as the preservation of habitat diversity, and others quasi-public goods, such as countryside use for hunting, viewing or exercise. Importantly, the outputs are often multiple and jointly produced with private goods, which are priced in the market. The competition between environmental goods and private goods can be illustrated by the production possibility frontier G (Fig. 7.1). It bonds the set of all technologically feasible combinations (X,Q), where X is a vector of private agricultural commodities and Q is a vector of environmental goods. The trade-off between X and Q is easily explained in terms of the marginal rate of transformation (MRT): because of declining marginal productivity of resources used for either output, G is concave to the origin and MRT is increasing. With increasing Q, farmers must give up more and more X to produce one additional unit of Q. So the marginal cost of producing Q increases with Q (X being the numeraire).

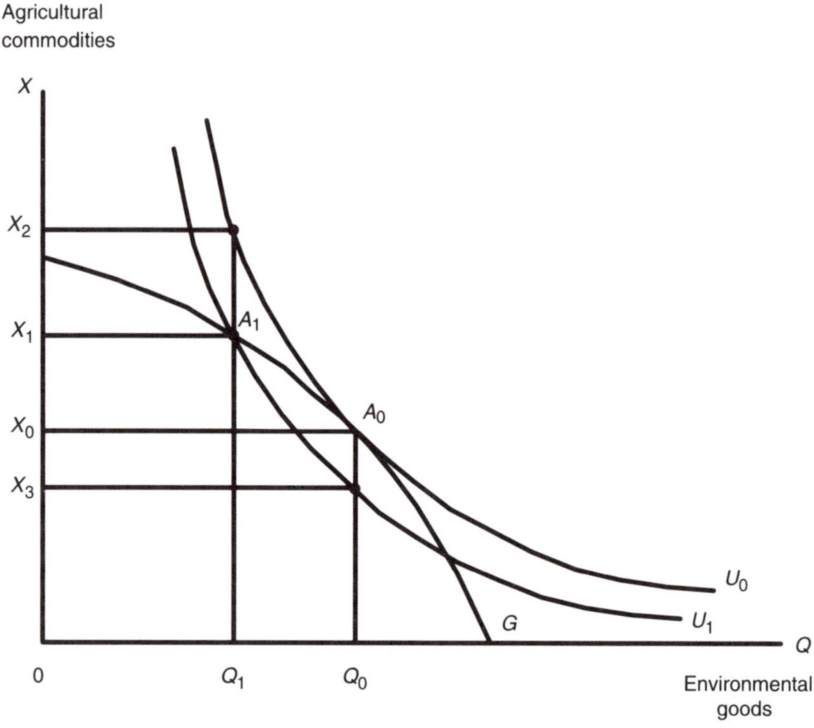

Fig. 7.1. Valuation of an increase in environmental goods.

Define social welfare as $U(X,Q)$. The Pareto efficient allocation is given by a feasible combination (X,Q) which maximizes social welfare. Graphically, there is a single solution, illustrated by A_0 which lies at the tangency of the isosocial curve U_0 and the production possibility frontier G. At the tangency, the exchange rate, defined in terms of social welfare, which is the marginal rate of substitution (MRS) between X and Q, is set equal to the exchange rate between the goods defined by technology (MRT).

Let us consider point A_1, which lies on G. Here society can only achieve a level of utility indicated by the social indifference curve, U_1, such that:

$$U_1 = U(X_1,Q_1) < U(X_0,Q_0) = U_0$$

In comparison with the optimal situation, there is an overproduction of agricultural commodities ($X_1 > X_0$) and an underproduction of environmental goods ($Q_1 < Q_0$). There are several strategies to establish the value of an increase in Q from Q_1 to Q_0 (Bonnieux and Weaver, 1996).

A first strategy is based on the supply side. At point A_1, farmers must give up X to produce more Q at a rate defined by the slope (MRT) of the production possibility frontier G. Consider a change from Q_1 to Q_0, then $\Delta Y = X_1 - X_0$

indicates how much X would have to be paid to farmers to produce Q_0. ΔY gives the profit foregone and therefore the minimum compensation to be given to farmers. This approach has been considered in the NRP context (Bonnieux *et al.*, 1995).

A second strategy relies on the demand side. It uses the various approaches of the consumer surplus. Basically, there are two possibilities in pricing an increase in Q. The first is to consider the compensating surplus (CS), such that:

$$U(X_1, Q_1) = U(X_3, Q_0) = U(X_1 - \mathrm{CS}, Q_0)$$

and

$$\mathrm{CS} = X_1 - X_3$$

The second measure is based on the equivalent surplus (ES), defined by:

$$U(X_0, Q_0) = U(X_2, Q_1) = U(X_0 + \mathrm{ES}, Q_1)$$

and

$$\mathrm{ES} = X_2 - X_0$$

The choice between CS and ES depends on the entitlement of property rights. Compensating surplus implies property rights in the status quo, so the welfare reference level is given by U_1. Equivalent surplus implies property rights in a state of change and so U_0 gives the welfare reference level. Under standard curvature properties shown in Fig. 7.1:

$$\mathrm{MRT} \leq \mathrm{MRS} \text{ for } Q \leq Q_0$$
$$\Delta Y < \mathrm{CS} \leq \mathrm{ES}$$

The equality holds if there is no income effect.

The CS is a measure of willingness to pay (WTP) to obtain an increase in Q. Otherwise, the equivalent surplus is a measure of willingness to accept compensation (WTA) to forego an increase in Q. Therefore, as $Q \leq Q_0$:

$$\Delta Y < \mathrm{WTP} \leq \mathrm{WTA}$$

An increase in Q is valued more highly by consumers than by the farmers that produce it. When there is no income effect, WTP and WTA are theoretically equivalent. However, when both are measured empirically, WTA tends to exceed WTP, even when income effects are insignificant. In this chapter, a conservative approach is followed and WTP measures are relied on (Arrow *et al.*, 1993). A third strategy could be based on a surrogate market. Graphically, changes in Q are valued along the tangency of the isosocial welfare curve U_0 and the production possibility frontier G at point A_0. This underlies the hedonic approach, which has been explored in the same context (Le Goffe, 1996).

The most direct benefits of the project are derived by visitors and those households living in or near the NRP. However, non-use benefits could concern non-visitors, who, for altruistic motives, are willing to pay to restore the *bocage*. The landscape of the NRP has many close substitutes in several locations of western France and the area of interest is limited. There is a small number of visitors, so the valuation (CV) has been restricted to the resident population. The WTP for the project is derived from a contingent valuation (CP) survey of the residents (Table 7.1).

THE CONTINGENT VALUATION SURVEY

In comparison with CV studies in the field of recreation, the application of the CV method (CVM) to rural landscapes is more subject to misspecification of the product. In particular, geographical part–whole bias and policy-package part–whole bias are likely to result from inadequate specification of the amenity. Geographical part–whole bias occurs when a respondent values a landscape whose spatial attributes are larger or smaller than the spatial attributes of the researcher's intended landscape. This issue has received special attention in our research. People were asked to value the whole area affected by Dutch elm disease; thus the valuation question refers to a regional landscape. In addition, questions were related to smaller areas, in order to assess differences in WTP. There was empirical evidence that people are able to distinguish a local landscape from a regional landscape. Respondents were able to isolate the regional landscape from a smaller area covering the specific village in which they were living.

Policy-package part–whole bias occurs where a respondent values a broader or a narrower policy package than the one intended by the researcher. This issue is of particular interest within the context of our research. Given that agricultural activities and the restoration programme can produce a wide variety of public effects, the public good to be valued had to be precisely specified. In the context of the Common Agricultural Policy reform, people might have considered the restoration scheme as symbolic of a larger policy package (for example, preservation of traditional farming) and assigned to the scheme some of the values they gave to this more general goal.

The cost of restoration should be supported by residents through local taxes. The population concerned includes both people living in the area to be restored and people living near this area. The survey was focused on a 400 km^2 area, in which there are 10,118 households. As most French households have a telephone, the phone directory was used to sample the population and no sampling-frame bias is assumed to have occurred. People were then asked by phone to meet an interviewer at home to participate in the survey. They were informed about the topics dealt with in the questionnaire and were told that the face-to-face interview would need no more than 30 min. Forty per cent of the contacted individuals refused to be interviewed, but some of them agreed to answer some simple questions. There is some evidence that refusals were more frequent among older people. The questionnaire was tested with a sample of 60 households in February 1995. The final sample includes 400 households. People were interviewed by technical staff, with a basic training in CV, during April and May 1995.

One contingent market was proposed to the interviewees. The corresponding good was constructed from the generalization of the pilot scheme over the whole area affected by the disease. The good was shown, using a combination of pictures and photographs. There was a detailed description of the system of standard agreements between farmers and the NRP, with an emphasis on the costs and on the way in which they are financed. One-third of the people were aware of this mechanism and 88% supported this environmental policy. This provides some evidence that the scenario is plausible. However, a significant proportion of the interviewees (30%) underlined inconsistencies of public policies. For example, they stated that the modern development of agriculture conflicted with countryside stewardship. So they were wondering about the possible integration of environmental goals into agricultural policy. Their comments show that people are well informed about environmental problems in the area of interest.

After a reading of the scenario, people were asked to reveal their WTP additional local taxes in order to carry out the restoration programme. The referendum question format was used: the respondents were asked how they would vote if faced with a restoration programme in exchange for higher taxes. Here the crucial issue is the choice of the bids offered to the interviewees, since it affects the estimation of mean WTP (Duffield and Patterson, 1991; Cooper, 1993; Cooper and Loomis, 1993). This choice was based on the outcome of the pretest survey, in which an open-question format was considered. People were assigned randomly to six subsamples, with each subsample being asked to respond to a different amount. The following amounts were selected: 50,100, 150, 200, 300 and 400 FF; the highest bid equals the second highest value obtained in the pretest survey. People are equally distributed according to the yes and no responses (48% each) and 4% of the sample refused to participate in the valuation exercise. The follow-up question allows for the categorization of the no respondents. Seventy per cent judge that local taxes are too high and would prefer spending money on other goods. Thirty per cent consider that farmers are responsible for the degradation of the countryside so they are not willing to pay for them. Some people in the latter category might be considered as protest bidders; there is, however, some evidence of free-rider behaviour. Older people are more inclined to refuse the offer, but some of them would agree to pay for a local good, such as a scheme restricted to their own village. The exclusion of these answers results in a greater WTP value. They are taken into consideration in the calculations which are given in the rest of the chapter.

In order to convert yes or no responses to a referendum question into a monetary measure, we employ Hanemann's (1984) model of choice. Several models have been used and they give predictions which are around 80% correct. Among those variables explaining the probability of accepting the offer, the most significant and the most robust in passing from one model to another are the bid level, income, education, angling practice outside the local community and concern. The signs of coefficients are consistent with theoretical expectations, since bid level, income and education positively influence the probability. Males and people having a recreational experience locally are more inclined to accept the offer. People who feel concerned about the protection of the whole area are willing to pay more, as the sign of the concern variable shows. This is a good indication that geographical part–whole bias is not a serious problem in our research. Angling experience outside the village negatively influences the probability for accepting the offer. Because this variable is a proxy for recreational substitute sites, this is consistent with demand theory.

Table 7.2 provides yearly WTP per household. Median and mean estimates are derived from two versions of the model (for the mean, two truncations are considered); with the second version of the model, education is used as a proxy variable for monthly income. Results derived from both versions are very close.

On average, 200 FF is a good estimate of yearly WTP per household. This varies according to the values taken by the explanatory variables (Table 7.3). The income elasticity of the demand for restoration is 0.9, implying a difference in WTP equal to 300 FF between the extreme income intervals. People who are concerned about the protection of the whole area are willing to pay 110–120 FF more than people who are not. Recreational-experience variables significantly affect WTP; for anglers, there is a large difference in WTP between those who fish locally and those who fish outside the area.

Table 7.2. Welfare measures (FF per household per year).

WTP measure	Model 1	Model 2
Median	201	198
Mean (0–400 FF truncation)	201	199
Mean (0–1000 FF truncation)	234	227

Table 7.3. Influence of individual characteristics on median WTP (FF per household per year).

Variable	Model 1	Model 2
Sex	58	66
Monthly income	35	–
Education	–	24
Concernedness	122	111
Nature-watching locally	92	70
Angling locally	102	110
Angling outside	–161	–191

THE RESULTS OF THE COSTS AND BENEFITS ANALYSIS

Planting and regeneration costs equal 13,000 FF km^{-1} and 240 km are planned to be planted yearly over the first 10 years. Thus, overall planting and regeneration costs are:

$$C_1 = 7.92 \sum_{t=0}^{9} (1 + r)^{-t} \text{ million FF}$$

where r is the discount rate. Maintenance costs are linearly increasing from the first year until the tenth year, when they reach a plateau. As they equal 1000 FF km^{-1}, overall maintenance costs are:

$$C_2 = 0.24 \sum_{t=1}^{9} t(1 + r)^{-t} + 2.40 \sum_{t=10}^{99} (1 + r)^{-t} \text{ million FF}$$

Based on market prices, the value of firewood is given by:

$$B_1 = 5 \sum_{t=15}^{99} (1 + r)^{-t} \text{ million FF}$$

and the value of timber wood is:

$$B_2 = 2.92 \sum_{t=60}^{69} (1 + r)^{-t} \text{ million FF}$$

As benefits to farmers are proportional to the length of hedges, the positive effect on agricultural productivity is:

$$B_3 = 0.072 \sum_{t=1}^{9} t(1 + r)^{-t} + 0.72 \sum_{t=10}^{99} (1 + r)^{-t} \text{ million FF}$$

where the benefits stemming from 1 km equal 300 FF.

Public benefits (amenities and recreational benefits) are assumed to increase with the development of the project and to reach a plateau. The overall public benefits are equal to:

$$B_4 = 0.10\, n\text{WTP} \sum_{t=0}^{9} t(1 + r)(1 + r)^{-t} + n\text{WTP} \sum_{t=10}^{99} t(1 + r)^{-t} \text{ million FF}$$

where WTP is 200 FF per household and n is the number of households who benefit from the project. The calculation has been limited to the resident population, so n equals 10,118 households.

Therefore, the net present value (NPV) of the project is:

$$\text{NPV} = B_1 + B_2 + B_3 + B_4 - C_1 - C_2$$

Net present value is reported in Table 7.4 for different discount rates. With a 4.8% discount rate, benefits balance costs (Table 7.4). It is arguable that the internal rate of return (IRR) is significantly greater, because some benefits are not taken into account. In addition to the benefits considered in Table 7.4, two

Table 7.4. NPV of the project for various discount rates (million FF): 100-year time horizon.

	Discount rate (%)				
	3	4	4.8	5	6
Costs					
Planting and regenerating	69.59	66.81	64.72	64.21	61.79
Maintaining	66.00	49.38	40.38	38.54	31.07
Benefits					
Firewood	101.26	69.61	53.03	49.71	36.59
Timber wood	4.35	2.34	1.43	1.27	0.69
Agricultural productivity	19.80	14.81	12.11	11.56	9.32
Amenities, recreation	61.18	46.18	38.11	36.37	29.60
Net present value	51.00	16.75	−0.42	−3.84	−16.66

other categories deserve some attention. First of all, people crossing the area enjoy amenities from viewing the scenic landscape. As the average daily traffic equals 11,000 cars, there is some evidence that the corresponding benefits are significant. The second category stems from the tourism industry. In a broader context, a positive relationship between rural amenities and the rent for second homes has been found (Le Goffe, 1996). This result is based on a hedonic equation and can be transferred to the situation prevailing in the NRP. Taking into account these two categories, a 5% IRR could be defended. However, based on use values only, the social profitability of the project is limited.

The major part of the costs is borne during the first 10 years, whereas the benefits are spread over the whole life of the project (Table 7.5). Therefore, in order to enforce the beneficiary-pays principle, such a project should be financed by a public loan. Otherwise, a significant money transfer would benefit the next generation.

CONCLUDING COMMENTS

In the NRP context the economic appraisal of landscape restoration results in a 5% IRR. In comparison with other private projects, as well as public projects, this is quite a low rate. As public benefits are important, a study restricted to private benefits (firewood, timber and impacts on farm productivity) would have been of very limited interest. These private benefits are not high enough to justify such a restoration. So the main contribution of this chapter is to extend the scope of project appraisal by incorporating amenities and recreational values.

Table 7.5. Costs and benefits over the life of the project (million FF): 4.8% discount rate.

	Years					
	0–9	10–14	15–59	60–69	70–99	0–99
Costs						
Planting and regenerating	64.72					64.72
Maintaining	8.07	6.85	22.79	1.18	1.49	40.38
Benefits						
Firewood			47.48	2.45	3.10	53.03
Timber wood				1.43		1.43
Agricultural productivity	2.42	2.06	6.84	0.35	0.44	12.11
Amenities, recreation	9.03	6.17	20.51	1.06	1.34	38.11
Net present value	–61.34	1.38	52.04	4.11	3.39	–0.42

As regards the valuation of public benefits, two points must be made. First of all, the long-running decline in hedgerows (for example, from 70 m ha^{-1} in 1975 to 27 m ha^{-1} in 1987) is expected to continue in the near future, resulting in a significant decrease in the supply of amenities and recreational opportunities. There is clear evidence that the demand for amenity and recreation is income-elastic (Fisher and Krutilla, 1985). A WTP–income elasticity equal to 0.9 has been found in this study, which should lead to higher values for WTP in response to higher incomes in the future. The combination of a downwards shift in supply with an upwards shift in demand involves a greater profitability of the project.

The process of aggregation used to derive the NPV of the project relies on strict assumptions regarding social welfare orderings (Boadway and Bruce, 1984). The marginal utility of income is assumed to be constant across the population of interest; this implicitly means that the income distribution is optimal. As the resident population of the NRP is poorer than the average population of France, it is arguable, based on equity considerations, that the NPV should be calculated with higher weights for low-income people. This alternative process of aggregation would lead to a greater value of the IRR.

As non-use value is not significant, the CV approach is relevant for valuing non-marketed goods (Diamond and Hausman, 1994). The degradation which occurs in the area considered is not irreversible and the concerned population is clearly identified and limited in scope. Thus, altruistic motives are expected to be marginal and the resident population to behave hedonistically. This also supports the use of the CVM (Enneking and Rauschmayer, 1996). The issue of part–whole bias remains a crucial one and has received attention in this chapter. Moreover, an external test of validity could be based on other studies. Here it may be emphasized that the CV results are consistent with the outcome of a hedonic study (Le Goffe, 1996).

NOTE

1 1995 ECU = 6.47 FF.

The Full Cost of Stewardship Policies

8

Martin Whitby, Caroline Saunders and Christopher Ray[1]

INTRODUCTION

Stewardship policies, aimed at encouraging the production of public environmental goods, often jointly with agricultural and forestry land use (see Weaver, Chapter 11, this volume), have been an important element in the policies of European Union (EU) member states for some decades. During the past decade, starting with Council Regulations (European Economic Community (EEC)) 797/85 of 30 March 1985 and 1760/87 of 15 June 1987, the EU has begun to promote such policies, using financial incentives. The policies were further extended through the measures accompanying Common Agricultural Policy (CAP) reform, in particular Council Regulation (EEC) 2078/92 of June 1992. They have since been consolidated in Council Regulation (EEC) 746/96 of 24 April 1996, which particularly emphasizes monitoring and evaluation (Articles 16–20).

The main exchequer cost of such policies arises from the requirement to compensate participating farmers for any loss they may sustain as a result. Under current regulations, payments to farmers are reimbursed from the Guarantee Section of FEOGA at the rate of 75% in Objective 1 regions and 50% in other areas. This description implies three different economic agents who bear policy costs: farmers, member states and the EU.

The main policy mechanism used under this regime is the management contract between farmers and the state, whereby they accept constraints on their business in return for compensation. This voluntary, and temporary, transfer of property rights requires information, contract or negotiation and enforcement or policing, elements which are defined by Bromley (1991) to

 The Economics of Landscape and Wildlife Conservation (eds S. Dabbert, A. Dubgaard, L. Slangen and M. Whitby)

constitute transactions costs (TC). Such costs will necessarily be incurred by all three agents. Farmers have to acquire information about the policies and analyse their implications, enter into contracts and negotiate and collect compensatory payments. States must set up the policies, defining where they apply and what they require, selecting and making rates of payment and monitoring compliance with contracts made; enforcement is also required. In the case of stewardship policies, the incidence of such costs among the three agents is typically not well documented. The levels of TC incurred by farmers are rarely measured; those involved at the state level may well be lost in public finance systems designed without promotion of efficient allocation of administrative resources in mind and at the EU level, where regulations are drafted and part of the cost must be paid by member states.

Article 9 of Council Regulation (EEC) 746/96 explicitly states that: 'The costs of preparing aid applications shall not be taken into consideration when determining the level of the aid.' A subsequent article makes it clear that member states are expected to verify compliance and undertake monitoring activities. The EU thus bears the costs of initiating such policies and some administrative burdens associated with ensuring payments to member states. But member states are left with the majority of the administrative costs of the policy and particularly with the responsibility of dealing with, typically, some thousands of farmer participants.

It is the central argument of this chapter that TC absorb substantial resources and that these amounts should feature in the analysis of policy costs. Although such arguments receive support in the economic literature (for example, Mathews, 1986; Wallis and North, 1986), it is unusual to find policy evaluations which include them. The chapter now details some evidence from the UK on the exchequer cost of transactions.

THE EXCHEQUER COST OF TRANSACTIONS

For most contract-based policies, it is believed that the majority of TC are incurred by public agencies. There is fragmentary evidence in support of this from one UK stewardship policy applying in Sites of Special Scientific Interest (SSSI), published by Whitby and Saunders (1996). Under this policy, individual contracts are negotiated with farmers, through which they are reimbursed for the farm income forgone in compliance and their negotiating costs are returned to them by the state. The system runs the obvious risk of overcompensating farmers if they can convince the authorities that their costs are larger than they actually experience. This may be a particular risk with negotiating costs, where norms may not be well established, compared with the standard costs that are available for agricultural operations. For the 1 year for which such data are available, the distribution of annual expenditure is detailed in Table 8.1.

Table 8.1. Estimated distribution of public expenditure on SSSI (1988/89) (from Whitby and Saunders, 1996).

Expenditure heading	Per cent of total
Cost of compensation	53.8
Landowners' negotiating costs	2.4
Other public expenditure	43.8
Total	100.0

The cost of compensation is based on the concept of income forgone, which should include the loss of income as a result of reduced farming intensity and any management and other costs of compliance with the agreement. The landowners' negotiating costs here would typically include professional advice from lawyers and estate agents. The other items of public expenditure include the professional advice obtained by the conservation agency and its staff and overhead costs, as well as the cost of designating SSSI. The presentation underlines the importance of TC in this case and sheds some light on the comparatively small negotiating costs initially borne by farmers.

The incidence of Bromley's (1991) three elements of TC – information, contracting and policing – varies in their shares of TC over the development of policy instruments. Applying his scheme to typical state–farmer contracts would suggest that information costs will be important in setting up the contract situation – defining which type of land is subject to the policy and what activities are to be encouraged or discouraged. Such costs are more likely to be of the 'overhead' type, incurred before the policy applied and spread over as much policy action as is taken. Initiation costs will then be incurred in attracting applicants to participate in schemes: informing them of their opportunities and encouraging participation from appropriate farmers. Contracting costs are likely to be incurred in advance of making an agreement; they would include surveying the area under negotiation, legal advice and negotiation costs. Policing costs are all those costs needed to ensure that the parties to the contract honour their commitment to each other. They are likely to be incurred throughout the life of the contract, perhaps becoming more important as the policy matures and benefits should begin to appear, and will include the costs of monitoring compliance.

The definition of TC stated above gives some indication of the problems to be expected in measuring these amounts. Transaction costs incurred by the state are likely to arise from activities in several separate agencies in the public sector. In practice, the English data relating to Council Regulation (EEC) 2078/92 come from the Ministry of Agriculture, Fisheries and Food (MAFF), which does nevertheless record the payments it makes to some other agencies for their work in helping to administer and monitor the policies. Data for

Table 8.2. UK distribution of annual expenditure on Council Regulation (EEC) 2078/92 (per cent of total) (from Agriculture Committee, 1997).

Cost heading	1992/93	1993/94	1994/95	1995/96
Payments	53.8	63.2	65.4	72.7
Running costs	33.7	26.1	25.4	19.8
Monitoring costs	12.5	10.7	9.2	7.5
Total	100.0	100.0	100.0	100.0

England for all policies implemented under Council Regulation (EEC) 2078/92 are recorded in Table 8.2 for the last 5 years.

This period began with a substantial expansion of activity as new instruments were set up under Council Regulation (EEC) 2078/92. Thus the shares of administration and monitoring, which are recognized here as the only data available on TC, reached more than half of the recorded costs in 1992/93 and declined to less than 25% of the total throughout the period. This trajectory emphasizes the importance of TC as new policies are being established.

A further element of potential cost may arise due to information asymmetry, where one side in contract negotiation succeeds in misleading the other into paying more than is strictly necessary (or accepting less than is due) to secure the contract. The process of negotiation is open to such deviations from the 'true' compensation needed to complete the contract. Where such events induce such a leakage of welfare from one party (probably the state, in this case) to another, they should be included as a cost, although, as long as they did not induce any reallocation of resources, they would not constitute a social cost.

Aggregate data for the costs of administering and monitoring SSSI and Council Regulation (EEC) 2078/92 policies were presented earlier, but those data suffer from two forms of asymmetry. First, under management contracts, such as Environmentally Sensitive Areas (ESAs), they are incomplete, in that they exclude private TC elements incurred by the contractee. For example, some ESAs require farmers to have a conservation plan drawn up, at their expense, before contracts may be made; other uncompensated legal and surveying costs may also be incurred. However, in the case of SSSI agreements, such costs were small in relation to total expenditure (Table 8.1); moreover, because ESAs use much simpler standardized contracts, their omission from the ESA cost data may be of small importance.

A second, more difficult, complication arises from the nature of the transaction. If each participant may gain from concealing information from the other, there is likely to be asymmetry of information between the participants and an optimal allocation of resources becomes unlikely. This will allow transfers between the parties, which may be important. For example, moral-hazard problems will arise if a farmer claims that his/her loss from participation in a contract is larger than he/she expects it to be, allowing him/her to extract more compensation than is strictly due. There will also be policy failures due to adverse selection, whereby participants might enter their least vulnerable land into the agreement, keeping the more vulnerable out of the agreement, perhaps in the hope of obtaining higher compensation at a later date. Equally, if a public servant successfully persuades the farmer to accept lower compensation than the instrument implies, perhaps in the cause of budgetary prudence, then he/she extracts a rent from the system, which, in this case, would be a social gain (to the UK government) to the extent that the alternative policy situation allowed the farmer to 'exploit' society. Whether such transfers through negotiation should be simply counted as benefits or as costs and the problems of measuring them remain difficult questions. If such mistaken transfers lead to changes in the level of taxes (as described by Oskam and Slangen, Chapter 9, this volume), they may also have important resource-allocation implications.

This section has shown that the exchequer cost of transacting stewardship policies can be substantial, especially during the establishment phases of a policy instrument. With rare exceptions, there is little information available on the private TC borne by producers. However, the reported costs are gross, in the sense that they include all costs associated with the stewardship policy, and in the next sections we consider the possibility of allowing for offsets to other policy costs resulting from stewardship. We now turn to the question of the social opportunity cost (SOC) of stewardship policies, as observable through ESA policies in England.

TOWARDS THE SOCIAL OPPORTUNITY COST OF ENVIRONMENTALLY SENSITIVE AREAS

The main stewardship policy in Britain was introduced, under Article 19 of Council Regulation (EEC) 797/85, in ESAs. There are 22 of these currently operating in England and their results have been comparatively well documented. By adversely modifying the allocation of resources in society, most policies reduce the welfare obtainable from available resources. Such losses are termed 'social opportunity costs'. In order to estimate SOCs of such policies, it is therefore necessary to know what effect they have on the use of resources, compared with a reference situation in which the policy did not exist. Here, this is attempted from an assessment of the impact of the first ESAs on agricultural output and the SOC of the transactions they generate. The steps in the argument must thus answer three questions:

1. What net effect do the policies have on farm resource allocation?
2. What demands do they make on the transacting sector?
3. What are the unit SOCs of these resources?

Farm Resource Allocation

As a case-study, we turn to the specific case of the ESA scheme and consider its net effect on exchequer costs and SOCs. This is done by reference to the land-use impact of ESAs and its estimated effect on output. From the available literature, the main output effects of ESAs are two. First, in several ESAs there is provision for arable reversion. This implies that land previously under crops may be returned to grassland as part of an ESA management contract. Where this happens, output may be affected in two ways: a reduction in crop output and an increase in grazing livestock output. Secondly, there is a general reduction in the intensity of grassland management, through reductions in fertilizer use, hay-cutting constraints and stocking-rate maxima. These effects must first be estimated in physical terms, before they can be valued using the unit values for SOC estimated above.

In the first ten ESAs designated in England, official monitoring studies were carried out by MAFF. The resulting reports (summarized in Whitby, 1994) give estimates of the effect of ESAs on land use and from their detailed findings an estimate of their impact on farm resource use has been prepared (Table 8.3). The MAFF studies did not, in fact, use a strict 'control' to identify the 'policy-off' or reference situation, so the effects they measured have to be interpreted with caution.

The reports relate to 1990 and 1991 at the end of the first 3 years of designation. The main effects they report are: the reversion of arable land to grassland, some changes in livestock numbers (both positive and negative at the level of the individual ESA) and substantial aggregate reductions in fertilizer use.

These effects are assumed to have been introduced during the 3 years of the programme but to be sustainable for the duration of the agreements, up to 5 years in total. Moreover, they are taken to be entirely policy-induced, because they are consistent with scheme objectives and because there are no data available for the 'policy-off' situation. After this period, some effects may have continued, although there are no data on this effect. Not all of the impact will have occurred immediately, so that these estimates relate to the end of the period of study, rather than the beginning. It is also very likely that further

Table 8.3. Estimated impact of ten English ESAs on farm output (1990/91) (from MAFF Monitoring Studies 1–10, summarized in Whitby, 1994).

	Size of impact
Suckler beef (LU)	2824.7
Hill sheep (LU)	136.0
Arable land (ha)	−5543.0
Fertilizer (t)	−1427.2

impact would have appeared by the end of the first 5 years of agreements, so these probably underestimate the fifth-year situation. Also not included here are data on net changes in capital, labour and materials inputs (apart from fertilizer) resulting from changes in agricultural practice. More important, there is no information on the physical requirements for increased environmental management activity. Although such activities are funded through the payments to farmers, their physical extent is not reported in detail in the monitoring studies.

Transacting Resources

The TC of policies under Council Regulation (EEC) 2078/92 were reported in Table 8.2 in terms of public expenditure. Those data could be divided by a notional cost per official to transform them to the physical measure called for here. But that would not contribute towards measuring SOCs. It would be more appropriate to estimate the net incidence of these costs, after allowing for policy offsets due to reductions elsewhere in the system, and then convert the net exchequer expenditures to SOCs, using a shadow wage rate.

The data for ESAs are reported as gross expenditures in Table 8.4. They will be converted to net expenditures after the next section, which presents unit exchequer costs for farm enterprises and allows estimation of the savings due to ESAs. The key conclusion from Table 8.4 is that the gross expenditure on transactions, to be carried forward and modified, was £10 million in 1994/95.

THE UNIT COST OF CHANGES IN AGRICULTURAL OUTPUT

This section examines the exchequer cost and SOC of agricultural output changes implicit in the reformed CAP. The changes here are marginal

Table 8.4. Exchequer expenditure on ESAs: England 1992/93–1996/97 (£ million, current prices) (from Hansard, House of Commons Debates, 10 December 1996, cols 107–101).

	1992/93	1993/94	1994/95	1995/96	1996/97 (estimate)
Payments	10,900	16,500	20,100	29,100	31,803
Running costs	7505	7968	7124	7053	5720
Monitoring	3603	4244	2871	3318	2266
Total	22,008	28,712	30,095	39,471	39,789

adjustments to the output in EU agriculture, at the member-state level, in that they do not affect equilibrium market prices for the commodities concerned; these changes affect a minute share of output through the implementation of ESAs. The results of these calculations are then used to estimate the costs of the ESA scheme.

There are two main stages in valuing changes in agricultural output. Firstly, the physical changes in agricultural output due to compliance with the stewardship policy constraints have been reported in Table 8.3. Secondly, the impact of these constraints on output has to be separated out to derive a cost–constraint matrix. The methodology for this chapter draws upon work examining the public exchequer and social cost of agricultural output lost (Saunders, 1996) and is used here to estimate SOCs.

There are three different values which may be relevant to any change in output: firstly, financial, that is, the change in the private costs and benefits of the farmer; secondly, public exchequer costs and benefits, that is, the effect of changes on exchequer payments; and, thirdly, social cost and benefits of a change in agricultural output. It is the latter two that are considered here. Henceforth, they are referred to as public exchequer costs and SOCs. The TC of these policies is reintroduced in calculating aggregate costs.

Public Exchequer Costs due to Changes in Agricultural Output

The level of exchequer support to the agricultural sector still accounts for a major share of total EU expenditure. In particular, since the 1992 CAP reforms, the level of public exchequer support in the form of direct payments has risen steeply. While such transfer payments should not be included in the SOC of agricultural output, they are clearly of interest to policy-makers and therefore cannot be ignored.

The Social Opportunity Costs of Agricultural Output

In valuing the SOC of agricultural output, the effect of government intervention on agricultural markets has to be removed. Market imperfections, which include imperfect market structures and unemployment of resources, should also be considered. But, due to the difficulty in assessing the impact of these various factors on the value of agricultural output, they have largely been ignored and perfect competition has been assumed in farm factor markets.

The basic system of support in the EU was, and to some extent still is, based upon the fixing of target, intervention and threshold prices. The intervention price is effectively a minimum price at which supplies are removed from the market by government agencies. The threshold price is the price at which imports are allowed into the domestic market and is maintained by a system of import levies. Other policy measures also continue to operate in the EU in addition to the above, including direct payments to farmers in Less Favoured Areas (LFAs).

There have been various reforms to the CAP, generally on a piecemeal basis, especially during the 1980s, but it is the Mac Sharry reforms in 1992 which have been the most comprehensive. While these left the basic price structure in place, they have reduced prices towards world market levels and compensated producers by direct payments based upon past production patterns. These changes are reviewed in detail later. Other policy measures also operate in the EU, such as quotas and direct payments, which, particularly as they are based on past farming performance, will affect resource use; the rights to transfer payments are frequently based upon past production patterns and therefore affect the distribution of farming types; they may be capitalized into the cost of inputs, especially land, and therefore affect resource allocation in the agricultural sector (especially in competition with other sectors).

Valuing the Social Opportunity Cost of Agricultural Output

There are two methods by which the SOC of the output from agricultural land can be determined. The first method is simply to revalue domestic output at world market prices, although this would overestimate the social cost of this output by ignoring the alternative uses of inputs and any protective measures on inputs. Secondly, social valuation can be undertaken by adapting effective protection rates (EPRs) to value the social cost of output from land. This would imply revaluing agricultural output at true world market prices on an undisturbed market and deducting from this the world market value of the inputs. The assumption that these inputs have alternative uses elsewhere is not always the case for factors such as capital, labour and other fixed costs in agriculture. Therefore, in calculating value added, it will be necessary to take into account the production systems used and the relative increase in output on farms. Further problems include allowing for fluctuations in exogenous factors, such as the weather, and so an average of EPR over a number of years is used. A full review of the methodological issues involved is given in Saunders (1996). Other problems include the estimation of protection on joint products or crops which are grown in rotation and therefore affect previous and future output; this is especially true with the introduction of set-aside. Strictly, the TC aspects of making CAP payments should also be included in all of these estimates, but data on such costs are not easily obtained and, on the assumption that these costs are rather small, they have been omitted. The next sections describe the public exchequer costs and SOCs of commodities relevant to ESAs.

Cereals

The CAP was, and still is, based upon price support, with the original cereal regime effectively forming the blueprint for other sectors. The policy involved the setting each year of institutional support prices as follows: a target price, an intervention price and a threshold price, maintained by import levies. However, with the Mac Sharry reforms, the only institutional price is now the intervention price, which has fallen in stages to be closer to world market levels. This was originally proposed to be 90 European currency units (ECU) t^{-1} in 1995/96, but, due to changes in the agri-monetary system, this has risen to 119.19 ECU t^{-1} for crop year 1995/96. The threshold price was abolished under the reforms and the import price is now equal to 155% of the intervention price, thus in 1995/96 it was 184.74 ECU t^{-1}. Export refunds still apply to exports, although their level is being reduced by 36% from the 1986–1990 level, over 6 years starting in 1995.

Producers are compensated for the fall in cereal prices by arable area payments (AAP), given the condition that they set aside a proportion of arable land, which is currently 5%. Small farmers, producing less than 92 t of cereals per year, are exempt from set-aside. The compensation farmers will receive is derived from the fixed tonnage payments, converted to area payments on the basis of regional yields. The compensation for set-aside is linked to the price compensation (57 ECU t^{-1}) for cereals.

The change in the cereal regime has effectively fixed the area and location of cereal production; thus any incentive to switch back to grass-based production, diversify or otherwise change farming practice has been removed. Lower cereal prices may encourage cereal-based feeding of stock, for example intensive beef production, but this will be counteracted by changes to the beef regime, outlined below. Thus the regime may well affect resource allocation, which should be considered in valuing the social cost of output. This would entail estimating the expected pattern of production, given no direct payments, so, for example, the relationship between farming type and relative prices could be estimated and applied to the current situation. However, as cereal prices have been maintained since the introduction of the reforms, relative to other sectors, it is unlikely that there would have been a major switch in production in the absence of the AAP, although this may well be possible in the future.

The public exchequer cost of the regime is therefore the AAP plus the export refund and storage costs of disposing of any surplus. The AAP for wheat and barley was £320 ha^{-1} for England in 1995/96. The average level of market support is calculated to be £3.48 t^{-1} or £26 ha^{-1}, giving a total public exchequer cost of £346 ha^{-1}.

The EU is self-sufficient in both wheat and barley, particularly in the case of feed-quality cereals. Therefore, the SOC of a marginal increase in production is the export price the UK can obtain on world markets. As the price in the UK is close to the export price, this has been used. Allowing for value added and the alternative use of inputs, the SOC of output was £314 ha^{-1} for wheat, £198 for winter barley and £130 for spring barley.

Forage

An output and input of concern for this study is conserved forage. To measure the output of forage involves imputing a world price, since there is no world trade in this product. Of possible ways that non-tradable goods can be valued, the method used here is to value forage output by converting the forage into its barley equivalent and using barley prices on the world market as a measure of the SOC of its output.

Sheepmeat

Support for sheepmeat is based upon an annual ewe premium, calculated as the difference between a representative market price and fixed basic price. The Mac Sharry reforms introduced rights to ewe premiums based upon historical production. The full ewe premium is payable on the first 1000 ewes in the LFAs and 500 in other areas, with any ewes above this number receiving only 50% of the rates of payment. Farmers in LFAs are entitled to an additional annual payment per ewe: the Hill Livestock Compensatory Allowance (HLCA).

Trade in sheepmeat is restricted by voluntary import quotas with the major non-EU suppliers. The public exchequer cost of sheep production therefore includes ewe premiums plus HLCAs. The ewe premium was an average of £19.65 per ewe in the lowlands and £25.24 on the hills. The HLCA was £5.75 per ewe at the higher rate, £3.00 at the lower rate, in the Severely Disadvantaged Area (SDA) and £2.65 in the Disadvantaged Area (DA). These public exchequer costs convert to £179 per livestock unit (LSU)[2] in the lowlands and £517 on hill farms.

The SOC of sheepmeat production is assumed to be the import price of New Zealand lamb, where most EU imports come from. As shown in Table 8.5, the SOC of hill ewe production is £153 LSU^{-1}, a similar amount to the SOC of lowland ewe production (Saunders, 1996).

Table 8.5. Exchequer cost and SOCs of unit changes in agricultural output (1994/95 prices) (from Saunders, 1996).

	Wheat (£ ha^{-1})	Barley (£ ha^{-1})	Hill ewe (£ LSU^{-1})	Suckler cow (£ LSU^{-1})
Exchequer cost	346	346	517	255
Social opportunity cost	314	165	153	–95

Beef

The CAP for beef is similar to that for the original cereals regime. A guide price is fixed and the intervention price is set at 90% of it. Intervention occurs when the market or reference price is below the intervention price for 2 weeks, and continues for 3 weeks after the reference price is above the intervention price. The changes to the beef regime include a 15% reduction in the intervention price over 3 years. The ceiling on intervention was to fall from 750,000 t in 1993 to 350,000 in 1997, with a new safety net if the market price falls to 60% of the intervention price.

To compensate farmers for these changes, headage payments on beef animals and suckler cows have been increased. The Special Beef Premium is not only raised but was originally to be paid twice in the animal's lifetime, at the ages of 10 and 22 months, although the Commission proposes to drop the second payment. The payment is subject to a maximum of 90 animals per holding and a stocking restriction.

The Suckler Cow Premium has also been increased but subject to a system of rights based upon previous stocking. The beef and suckler-cow premiums are subject to a maximum stocking density of 2 LSU per forage hectare. In calculating the stocking density, all animals eligible for premium are included, that is, ewes and suckler cows. There is an additional payment of 30 ECU per head if the stocking rate is below 1.4 ha^{-1}. As stocking densities are calculated using forage area, intensive beef producers lose their right to premiums beyond 1992, unless, of course, they have spare forage area.

Again, the introduction of rights to some extent freezes production patterns; however, the ability to sell and transfer rights to ewe and suckler cow premiums does reduce this to some extent. Beef cows in the LFAs are eligible for the HLCAs.

Therefore, the public exchequer cost of beef production includes the Suckler Cow Premium and the Special Beef Premium, as well as the cost of intervention buying and export refunds and any HLCA. The Suckler Cow Premium was £114 per cow and the Special Beef Premium £86, with an additional £29 extensification premium. The HLCA was £47.7 in the SDA and £23.75 in the DA (FBS, 1996). Thus, in hill land, the Suckler Cow Premium would be full rate plus the higher HLCA, giving a public exchequer cost of £255 LSU^{-1}. To estimate the SOC, beef production output has been revalued, using EU export prices of beef. The value added at world prices is –£95 LSU^{-1} on hill farms, implying a benefit to society of reducing output from cows.

Table 8.5 summarizes the unit values derived from the sections above. It indicates the unit value changes in exchequer cost and SOC that are brought about as a function of CAP-reform changes in agricultural output.

VALUING THE IMPACT OF ENVIRONMENTALLY SENSITIVE AREAS

The unit exchequer costs and SOCs are taken from Table 8.5 and are multiplied by the physical magnitudes of impact, reported in Table 8.3. The unit values for beef and sheep are expressed 'per LSU', which renders them equivalent to one dairy cow in terms of land use. To allow for the fact that ESA-agreement land is likely to be of low productivity, the resource value for arable land is taken as 80% of the arithmetic mean of average values estimated by Saunders (1996) for wheat and winter and spring barley. The effect of these calculations is to indicate a reduction in the gross exchequer cost of payments of £1.13 million and a SOC of £1.38 million. These adjustments apply to roughly half of the ESAs operating in England during the period 1989–1991, when the exchequer costs of payments to farmers in England was in the order of £10–12 million (MAFF, 1991).

Extending the estimates to the rest of the UK would require increasing the area covered: direct calculations are not available in the absence of monitoring information about other ESAs. Because the TC data available relate to England in 1994/95 the estimated impact on output must be projected forward to the same year. The area of ESAs in 1994/95 was more than twice that of 1991/92 and cumulative rates of uptake rather higher, so the estimates in Table 8.6 are raised by a factor of three, giving estimates for exchequer costs in the range –£3.4 million; the result is displayed in Table 8.7.

The reduction in exchequer costs occurs because interactions between ESA and other policies have been removed, leaving net exchequer costs which are smaller than the gross direct expenditures on ESAs. There might also be some reduction in TC associated with the reduction in output-related expenditures, but these are unknown in size and are taken to be small enough to

Table 8.6. Estimated impact of ESA designation on the cost of agricultural output: England, 1990/91 (from MAFF Monitoring Studies 1–10, 1991, 1992, and Saunders, 1996).

	Size of impact (LSU, ha or tonnes)	Exchequer cost change (£ per LSU ha or tonne)	Resource cost change (£ per LSU, ha or tonne)	Aggregate change in exchequer cost (£ million)	Aggregate change in SOC (£ million)
Suckler beef	2824.7	255.0	–95.5	0.72	–0.27
Hill sheep	136.0	516.7	283.3	0.07	0.02
Arable land	–5543.0	346.0	234.3	–1.92	–0.95
Fertilizer	–1427.2	0.0	126.1	0.0	–0.18
Total				–1.13	–1.38

Table 8.7. Estimated gross and net exchequer cost of English ESAs (1994/95).

	Gross exchequer costs (£ million)	Estimated change due to ESAs (£ million)	Total net pre-Fontainebleau exchequer costs (£ million)
Payments to farmers	20.10	−3.39	16.71
Transactions costs	10.00		10.00
Total	30.10	−3.39	26.71

ignore. Transactions costs are reported in Table 8.7, together with the other exchequer costs.

Because of the financial solidarity of the EU, expenditures reimbursed from central EU funds cost member states no more than the marginal amount of their contribution to the common budget. So, if a member state contributes 10% of the EU budget, the cost to its exchequer and hence to its taxpayers is only 10% of what is returned from the budget in Brussels. The position is further complicated in that, under the Fontainebleau Accord, the UK receives an *ex post* refund from the budget of two-thirds of the excess of its budgetary contribution over its receipts from the budget in any year. Such payments are necessarily delayed until the relevant amounts are known.

For the 1994/95 payments in Table 8.7, the sequence of calculations to reach the UK net exchequer expenditure is as follows:

		£ million
[1]	Expenditure net of output savings	16.71
[2]	European Agricultural Guidance and Guarantee Fund (EAGGF) reimbursement (27% of [1])	−4.51
[3]	UK contribution to EAGGF (15% of [2])	0.68
[4]	Fontainebleau abatement (66% of −[2]−[3])	2.53
[5]	Net UK exchequer expenditure	15.41

This sequence of calculations follows that of MAFF (1996). It further reduces the actual cost of payments made to British farmers by the UK government to £15.41 million, compared with recorded TC amounting to £10.0 million.

In Table 8.8 the SOC of ESAs in England is summarized. Here only the effects related to ESAs are allowed for, that is, the change in SOCs of output they induce and the TC they impose. The level of TC at £10.00 million reported in Table 8.7 may be greater than the SOC of transactions associated with the policy, reflecting the possibility that those employed in the transacting sector might not all be able to find alternative employment. To allow for this, TC is adjusted, arbitrarily, reducing it by 20%.

Table 8.8. SOC of ESAs in England (1994/95).

	Aggregate SOC (£ million)	Cost per hectare under contract (£)
Payments to farmers	–4.13	–13.1
Transactions costs	8.00	25.4
Total net social cost	3.87	12.3

These arguments lead to the set of estimates in Table 8.8. The main finding that the SOCs of ESAs are negative, while the TC they bring are positive, is important in policy terms. Dividing the aggregates by the area under agreement in 1995, in the second column, shows that management of each hectare under agreement costs £25 in SOC terms, whereas payments to farmers result in a net gain of welfare to society of £13.10 ha^{-1} (without any valuation of the environmental benefits obtained), giving an estimated net social cost per hectare of £12.30.

CONCLUSION

While the extreme crudity of the above estimates is acknowledged, they are by no means incredible, and it is hoped they will provide an incentive for governments to strive for transparency as to the level of TC and for economists to make more use of this information in evaluating policies.

What this means for policy-makers is an interesting area for speculation. At present, EU agri-environmental policies, as funded through Council Regulation (EEC) 2078/92, provide funding of compensation costs for member states at rates varying from 50 to 75%. The complex effect of the EU contribution to costs, as modified by the Fontainebleau Accord (which applies in the UK only), reduces the exchequer costs estimated. These results confirm that ESAs cost less to the UK government than the exchequer cost of the payments made to farmers, because of the savings induced in payments through other policies and the contributions from the EU budget. The net effect of this is to emphasize the importance of TC. In terms of SOCs, ESAs bring some small gains, through resource savings, to set against the substantial volume of transactions effort they require.

Yet, generalizing from the UK example, the negative SOCs of ESAs constitute a small positive benefit to member states, compared with the cost of the relevant transactions, most of which is borne at the state level and which, in this example, cost more to the UK government than the savings in SOC by reducing farm output through payments to farmers. Under such circumstances, it may be asked, 'What should a rational government do?' They could decline to participate in this regulation or perhaps lobby the Commission to meet some of their TC as well as the compensation costs. These estimates, extrapolated to the whole EU, would imply that the agri-environment budget of the EU, currently set at some 2 billion (2000 million) ECU, generates much greater expenditure in individual member states. The Commission may, however, require more evidence than this single case-study for the UK before making major changes in their behaviour.

In conclusion, it is evident that the full cost of these environmental policies is difficult to measure. The financial costs are obscured by the administrative mechanisms which deliver the public goods, and the public finance system within the EU does not make for transparency in funding policies. Even if the payments to farmers are carefully reported, they must be adjusted to allow for offsets to the cost of other policy instruments. Typically, the TC of such policies are by no means fully revealed. Private TC are completely ignored and the public financial cost of transactions is only partly revealed – to the extent of expenditure on monitoring and 'administration'. Undoubtedly, there are many other costs which are similarly obscured, and it must therefore be accepted that the full cost of environmental policies and hence the cost-effectiveness of the allocation of these resources are likely to remain obscure. We must therefore recognize that information is not infinitely valuable and accept that incomplete information is a fact of life. Nevertheless, where information may be collected cheaply, governments have a strong interest in doing so if they wish to approach effective use of public funds and national resources.

NOTES

1 The authors have received useful comments on this chapter from Professor David Harvey and Dr Ben White. They take full responsibility for remaining errors.

2 Livestock units based on the dairy-cow equivalent are used here, where one cow is equivalent to 17 hill ewes or nine lowland ewes.

The Financial and Economic Consequences of a Wildlife Development and Conservation Plan: a Case-study for the Ecological Main Structure in The Netherlands

9

Arie Oskam and Louis Slangen[1]

INTRODUCTION

For many centuries, agricultural land in Europe has been increased at the cost of wilderness areas and forests. During recent decades, the demand for land used for infrastructure, housing and – later – recreation has occupied an increasing area share of the land (Kamminga *et al.*, 1993). After 1950, one can observe a large real income growth per capita in Europe. In agriculture, the rising wages induced labour-saving and production-enhancing techniques, which increased agricultural output and helped the dwindling number of farmers to achieve income levels comparable to those outside the sector. A high level of mechanization, intensification of land use and specialization at the farm and the regional level accompanied the changes in agriculture. As a result, there was a deterioration of wildlife and landscape and the quality of soil, water and air.

While agriculture experienced these developments, higher incomes increased the demand for wildlife and landscape, for leisure and opportunities for outdoor recreation. In other words, during a time when the supply of wildlife and landscape and areas for outdoor recreation decreased, the demand for those amenities actually increased.

In comparison with other European Union (EU) countries, in The Netherlands the tension between the interests of agriculture as user of the land and the interest of the conservation of wildlife and landscape appears to be more pervasive, for the following reasons. In the first place, Dutch agriculture uses the land very intensively. A large part of Dutch agriculture consists of intensive livestock farming and horticulture. In the second place, The Netherlands has

 The Economics of Landscape and Wildlife Conservation (eds S. Dabbert, A. Dubgaard, L. Slangen and M. Whitby)

by far the highest population density in Europe (about 450 inhabitants km^{-2} compared with 150 in the EU-12), but has relatively few forest, recreation and wildlife areas. Together, these comprise some 530,000 ha, or 13% of the total land area (LEI/CBS, 1996, pp. 20, 217).

More recently, there is an increasing interest in converting cultivated land to wildlife development areas. Examples of present wildlife conservation areas are now available which previously had a quite different function (agriculture, sea and so on). An interesting example consists of the Ecological Main Structure (EMS) in The Netherlands (MLNV, 1990, 1992), which is intended to connect existing wildlife areas and to increase their total size. Evaluating the pros and cons of this plan is considered to be a typical cost–benefit analysis. According to our information, no example is available in the literature of such a cost–benefit analysis.

This chapter is structured as follows. The next section contains two different basic philosophies: one derived from the ecological perspective and the other from an economic perspective. The two theories complement each other. The methodology of a financial and economic analysis is then discussed. The section on information used in the analysis in relation to agriculture, agribusiness and conversion and maintenance costs of the EMS deals with an important share (on the cost side) of the information to perform the analysis. Specific environmental effects, including more land in a wildlife conservation area, are then dealt with in the section on effects for the environment, wildlife and landscape of the EMS. The main intention of the EMS is to increase the size and impact (and therefore the value) of wildlife and landscape. We try to incorporate all relevant information in the section covering the total economic value of wildlife and landscape conservation areas. Because the EMS is a long-term plan, price developments are very important (see the section on price developments). The high financial burden is government-financed and the economic consequences of government financing receive attention in the section on taxes and the effects of government financing. The penultimate section (a model to calculate the consequences of the EMS under different assumptions) provides the results for the most relevant assumptions incorporated in the base alternative and some very limited sensitivity analysis. The final section discusses some uncertainties and provides conclusions.

THE LOCATION AND EXTENT OF WILDLIFE CONSERVATION AREAS: THE THEORY

The basic philosophy behind the EMS is related to the ecological-island theory of MacArthur and Wilson (of 1963), who stated that isolation of wildlife conservation areas reduces biodiversity and therefore the value of those areas (MLNV, 1990; van Asperen, 1994). The EMS has been targeted to strengthen the wildlife and landscape value of the existing conservation areas by adding new areas at strategic places. Because almost the whole additional area consists of land currently used in agriculture, the EMS will influence production in agriculture and related sectors. It is clear within this philosophy that ecological considerations dictate the location of the EMS. The island theory is related to the economies of scale and endogenous growth theories in economics (Dosi, 1988).

Another basic philosophy is concerned with optimal allocation of land use, which depends on location and rent-generating capacity (Barlow, 1972, p. 15). Alonso (1964) developed a concentric-zone model of land use, using the concept of rent-bid functions, which give the per hectare generated rent for each alternative type of land use. This approach is closely related to the well-known Von Thünen theory, with a concentric zonation of land uses, seen from the centre of a city. According to Lloyd and Dicken (1978, p. 44), the following relationships hold:

- Different options of land use determine land values through competitive bidding among users.
- Land values distribute land uses according to their ability to pay. This ability depends upon the rent level in relation to the location.
- The steeper rent curves capture the central locations: in other words, those products or processes which have most to gain by locating near the market or centre will outcompete alternative land use.

This implies that the level and slope (shape) of rent-bid functions – which might also represent valuations of public goods and/or recreation areas – determine the optimal spatial allocation. Based on the synthesis of the models of Von Thünen and Alonso, it is possible to develop a theory of optimal spatial allocation of recreation and wildlife areas (von Alvensleben, 1995, p. 321). Figure 9.1 shows an example of three rent-bid functions. Here recreation areas are near to the city. A flatter rent-bid function of agriculture leads to a location more on the periphery.

According to the theory of Von Thünen, the location of real wildlife areas, where existence value is most important, is at places with the lowest land rent, which means at the periphery. The valuations of real wildlife areas do not depend on transport costs: the rent-bid function is flat. A negative relation (to the distance from the centre) is possible: near to the centre the pressure on wildlife areas can increase through recreation and pollution. However, it is also reasonable to assume that the valuation of wildlife areas is negatively related to their share in land use, at least above a certain threshold (von Alvensleben, 1995, p. 233). Supposing a decreasing marginal utility of preserving additional landscape and wildlife areas and land rent is higher near to the city (or to the farm) than further away, there will be a rent-bid function for wildlife as shown in Fig. 9.2. On the basis of this approach and assuming a public demand, wildlife areas will be destined to zones at a greater distance from the

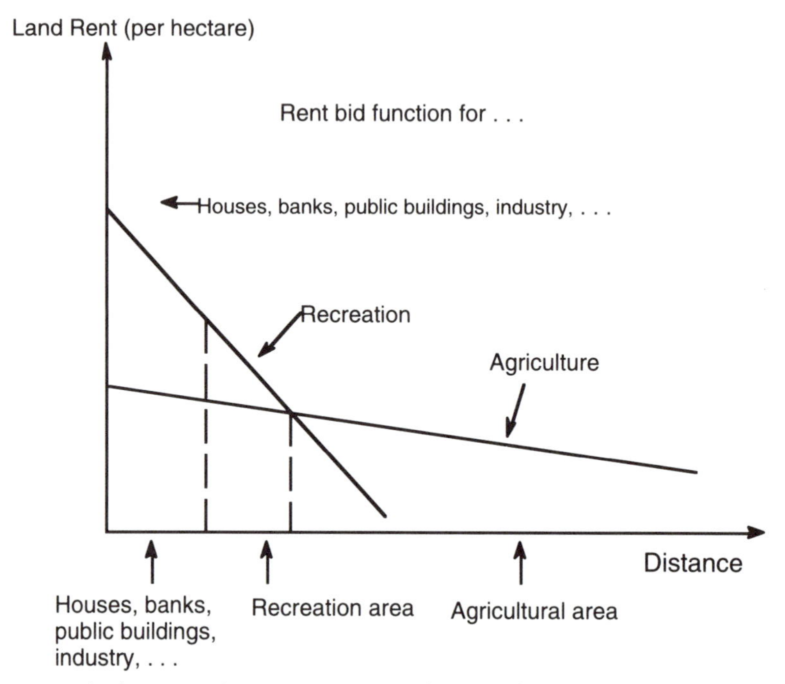

Fig. 9.1. The location of recreation areas in the Von Thünen–Alonso model.

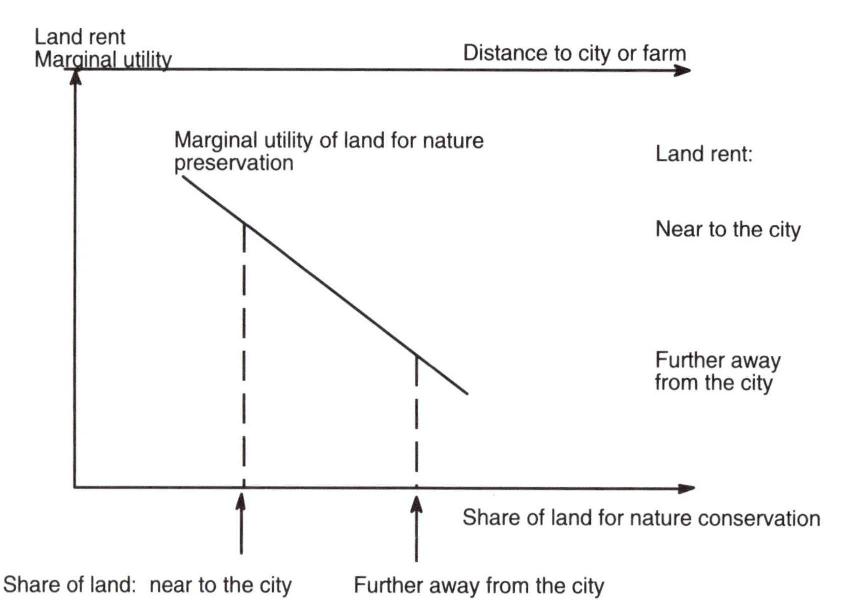

Fig. 9.2. Optimal share of nature areas.

city. The different land rents cause a larger share of wildlife areas at a greater distance from cities or markets.

The theoretical analysis of the optimal allocation of land for wildlife conservation purposes leads to the following conclusions (von Alvensleben, 1995, pp. 235–236):

- The optimal allocation of land for wildlife conservation is determined by the alternative land rent-bid function and the valuation of wildlife conservation.
- Given equal valuation of wildlife conservation at each location, the optimal share of land allocated for wildlife conservation is lower on land with high productivity than on less productive land.
- Planning that is only based on the ecological value of land and neglects alternative land rents is a deficient allocation method for land for wildlife conservation.

It will be clear that the optimal size and allocation of recreation and wildlife conservation areas require a lot of information, which is often not available. It is quite clear that, in designing the EMS, this methodology has not been used. Our analysis refrains from determining optimal size and allocation of the EMS. We take the EMS as given and make a cost–benefit analysis of the total project, reducing the rent-bid functions to some point estimates.

FINANCIAL AND ECONOMIC ANALYSIS: THE METHODOLOGY

Conversion of agricultural land to create wildlife areas has financial and economic consequences. Such a reallocation of the land use can be seen as a cost–benefits project. Here, a financial approach and a national-economic approach are distinguished. The financial approach considers the government as an agency; the project will be evaluated from the point of view of the government (state) budget. The terminology 'budget analysis' can also be used. The national-economic approach looks at the project from the perspective of society as a whole (Boadway and Bruce, 1984, p. 294). The general rules of cost–benefit analysis are followed from the perspective of The Netherlands, using opportunity-cost principles. Non-marketed goods will be valued in relation to the price of market goods. The production of market goods generates net added value, which is the measure of their value to the economy if prices reflect scarcities.

A financial analysis focuses on the revenues and expenditures of a particular actor – here, the national, regional or local government. There are two main reasons to include a financial analysis:

1. It provides an overview of the government budget costs of a project.
2. The government budget expenditure influences the economic process and therefore the economic effects of a project. Taxes influence economic decisions. According to Rosen (1995, pp. 303–325), they cause welfare costs or deadweight losses (for more details, see the section later in the chapter about taxes and the effects of government financing).

Table 9.1 gives an overview of entries that are important for the financial analysis. The financial deficit (FG) can be calculated if the other values are available. Table 9.2 provides the most important entries in the economic analysis, resulting in a total net cost of EH, which might also be a total net benefit if EE + EF + EG is larger than EA + EB + EC + ED.

An important issue is the valuation of the opportunity costs of wildlife conservation areas currently in agriculture. What would be the (forgone) net value added generated by the area that comes under the regime of wildlife conservation? In earlier studies, Slangen (1994a), Oskam (1994) and Sijtsma and Strijker (1995) tried to analyse this issue. None of them, however, has used the following short-cut or 'free land-price' approach. The free land price exactly reflects the present value of all future net benefits of a hectare of land that shifts

Table 9.1. Financial overview of converting agricultural land into a wildlife area.

Expenditure		Receipts	
Purchasing land	FA	Returns	FF
Planning and policy decision-making, administration	FB		
Cost of developing nature areas (reconstructing, planting)	FC		
Management, maintenance and operation costs	FD		
Reduced tax receipts	FE	Financial deficit	FG
Total	FT	Total	FT

Table 9.2. National economic costs and benefits of converting agricultural land into a wildlife area.

Costs		Benefits	
Loss of net production value in agriculture and agribusiness	EA	Net production produced by nature areas	EE
Costs of production factors required for nature preservation (developing, maintenance, etc.)	EB	Net production of production factors coming out of the agriculture and agribusiness	EF
Loss of emission capacity of agricultural land	EC	Non-market value of nature areas (compared with their original situation)	EG
Other costs (planning, administration, excess burden, etc.)	ED	Allocation losses	EH
Total	ET	Total	ET

to the EMS. The private net benefits are calculated from the perspective of the decision-maker, which is the farmer. Opportunity costs of labour, capital goods, future prices and quantities of outputs and variable inputs and also future land prices are considered in this approach. Even the value of land for spreading manure and the expectations on rules and regulations in land use, as well as expected length of quota regulations, are included in the price of land. This approach can be considered as a useful alternative method of calculating economic costs. We assume that the free land price equals the purchasing price. So, unless there are large distortions between private and public prices of inputs and outputs, the land price provides a good indication of the marginal value of land to be used in the EMS. Observe, in Table 9.3, that this method uses a mixture of financial and economic items. Allocation losses arise if FA + EB + ED is larger than EE + EK + EG.

INFORMATION USED IN THE ANALYSIS IN RELATION TO AGRICULTURE, AGRIBUSINESS AND CONVERSION AND MAINTENANCE COSTS OF THE ECOLOGICAL MAIN STRUCTURE

Determining the financial and economic consequences of the EMS requires a lot of information. Here, we rely on several pieces of information which have been determined in previous research (Jansen, 1994; Slangen, 1994a; Sijtsma and Strijker, 1995), on government papers (MLNV, 1990, 1992) or on recent research. Often choices had to be made, because of differences within the available information.

Our approach is summarized in the following details:

- A total of 150,000 ha will be withdrawn from agriculture over a period of 30 years (Loman, 1996, p. 19). This consists of 50,000 ha of wildlife development area, which includes Strategic Green Projects (SGPs), with mixed recreation and wildlife conservation objectives, and 100,000 ha of wildlife conservation area.[2]
- The weighted average land price is 40,500 Dutch florins (Dfl) ha^{-1}. This amount is based on voluntary sale by farmers. In the case of expropriation, the purchase price will be higher. Jansen (1994, pp. 26–31) has calculated the hectares per province to be purchased to realize the EMS. Based on De Schutter (1993, p. 26), the calculated weighted average price of sand, clay and peat land is about 40,500 Dfl ha^{-1} in The Netherlands. Loman (1996, p. 50) calculated an average land price in 1996 of 42,500 Dfl ha^{-1}.
- In our approach, it is assumed that only land in use as arable land and dairy farming will be converted to wildlife area. This involves activities which need agricultural land as an important production factor. The net value added from dairy farming was, during the period

1990/91–1994/95, on average 4410 Dfl ha^{-1} $year^{-1}$. In the same period the net value added generated by specialized arable farms was 2629 Dfl ha^{-1} $year^{-1}$. Weighted to their shares, the average net value added was 3750 Dfl ha^{-1} $year^{-1}$.

- Net agricultural production is realized by the production factors labour, capital and land. Labour and capital can (partly) be reallocated to other sectors of the economy, although returns might be lower. The economic importance of land within agriculture can be based on the marginal value or the shadow price of land (Table 9.4), which has been derived by differentiating the short-run profit function of land (see, for example, Helming *et al.*, 1993, p. 348). Land has a share of 47% of net added value. Converting 150,000 ha of agricultural land to wildlife areas means that farms will disappear and an average approach to valuation would be better. Shadow prices, mainly based on between-farm differences in profit levels because of different farm sizes, do not provide the most relevant information. In our calculation, we use a share of land rent in net added value of 30%.
- The net value added of the delivering and processing industries related to agriculture is nearly equal to the net value added of agricultural production (Peerlings, 1993). Here, a net value added of 3500 Dfl ha^{-1} $year^{-1}$ is

Table 9.3. National economic costs and benefits of developing wildlife areas, based on free land-price approach.

Costs		Benefits	
Free land price of agricultural land	FA	Net production produced by nature areas	EE
Costs of extra production factors required for nature preservation (developing, maintenance, etc.)	EB	Net production of production factors originally in agribusiness	EK
Other costs (planning, administration, excess burden, etc.)	ED	Non-market value of nature areas (compared with their original situation)	EG
		Allocation losses	EL
Total	ES	Total	ES

Table 9.4. Overview of estimated shadow prices of agricultural land.

Research	Reference year prices	Shadow price (Dfl ha^{-1} $year^{-1}$)	In prices of 1992/93
Helming *et al.*, 1993	1980/81	1119 (G)	1427
Helming *et al.*, 1995	1980/81	1348 (G)	1719
Oude Lansink, 1994	1980/81	1683 (A)	2146
Oude Lansink and Peerlings, 1996	1992/93	1968 (A)	1968
Boots *et al.*, 1997	1992/93	1629 (G)	1629

G, grass; A, arable.

assumed, because there are indications of a reduced agricultural production, which has – on average – a less strong relationship with upstream and downstream industries (cereals, green maize, sheep and beef farming and so on)

- The costs of converting agricultural land to 'new' wildlife equals about 7000 Dfl ha^{-1}.[3] This is based on a weighted average of several different types of wildlife, which includes the SGP (Grontmij, 1987; Jansen, 1994, p. 24; Loman, 1996, p. 32).
- We assume that purchased land on average takes 3 years before it is converted. The financial return has been put at 400 Dfl ha^{-1} $year^{-1}$ and the economic value is assumed to be 800 Dfl ha^{-1} $year^{-1}$.
- The maintenance costs of a wildlife conservation area amount to 1500 Dfl ha^{-1} $year^{-1}$.[4]
- The costs for water boards and polder boards are assumed to be 150 Dfl ha^{-1} $year^{-1}$.
- In 1994, 1 European currency unit (ECU) = 2.15 Dfl.

All this information will be included in our analysis.

It is not easy to calculate alternative revenue of labour and capital used in agriculture and agribusiness. Several different approaches might be used. It is quite reasonable to assume a short-term alternative revenue that is smaller than present net value added generated in agriculture and in the related agribusiness. Furthermore, opportunities of related sectors are quite often easier to realize than alternative destinations of labour and capital in agriculture. Production factors are more mobile, and related industries do not often depend completely on local agriculture. A rather long period of realization of the EMS makes it possible to buy a substantial part of the land from retiring farmers (without a successor). Moreover, the knowledge that land use in agriculture will decline because of the EMS makes it possible for related industry and also farmers to anticipate such changes.

On the basis of these arguments, we assume that in agriculture 50% of the labour and capital will have no alternative revenue in the first year. This declines stepwise to zero over a period of 10 years. For the related agribusiness we assume that 25% of the labour and capital receive no alternative revenue in the first year. This declines stepwise in a period of 5 years to zero. These assumptions imply long-term flexibility for all factors of production in the economy.

EFFECTS FOR ENVIRONMENT, WILDLIFE AND LANDSCAPE OF THE ECOLOGICAL MAIN STRUCTURE

Over a period of 30 years, the EMS results in an additional withdrawal of agricultural land of 150,000 ha. Given the rather intensive types of agriculture that exists in arable and dairy farming and taking into account that milk quota, sugar quota and potato growing will largely remain because of the high profitability of these products, an intensification of agricultural land use in the rest of the area might be expected. In the empirical analysis, a simple approach is followed to calculate the costs of reduced land use in agriculture. The equilibrium use of phosphate is for grassland 110, for maize land 75 and for arable land 70 kg $year^{-1}$. Emission rights are estimated to have a price of about 5 Dfl kg^{-1} $year^{-1}$.[5] A weighted average value per hectare of the emission rights is approximately 400 Dfl.

Several environmental aspects might be considered:

- The production of the minerals nitrogen (N), phosphorus (P) and potassium (K) caused by livestock will hardly change.
- Ammonia emissions will not change much; ammonia emissions are mainly related to dairy farming and intensive livestock farming (Mulder and Poppe, 1993), which will not change much.
- Reduced local emissions of nitrate, phosphate, K and pesticides to groundwater and surface water, due to the shifts in land use to wildlife conservation areas.
- A more intensive land use in other areas, with higher level of emissions of N, P and K and pesticides.
- A very small change of greenhouse-gas emissions (van Bergen and Biewenga, 1992; Simons *et al.*, 1994).

Summarizing this information, it seems justified to estimate no important real positive net effect of the EMS on the quality of soil, water and air in this respect.

Converting agricultural land into wildlife areas will sacrifice the non-marketable value of wildlife and landscape of agricultural areas. Except for reduced phosphate emissions, the effects of a more intensive use of the agricultural area – because of the decisions of farmers – on landscape and wildlife values and the quality of the still existing natural elements have not been quantified. In the next section, an overview of the total economic value of land use is provided.

THE TOTAL ECONOMIC VALUE OF WILDLIFE AND LANDSCAPE CONSERVATION AREAS

The main objective of the EMS is to create valuable wildlife and landscape conservation areas. Therefore, special attention is given to the valuation issue. Sijtsma and Strijker (1995) provide a lot of information on the wildlife valuation, both from a biological perspective and on preference valuation of different people.

The total economic value of land use includes marketable benefits, as well as non-marketable benefits. Natural (wildlife) values and landscape values of natural areas are a part of the non-marketable values. The value of an environmental asset like wildlife, natural elements and forestry can be measured by the preferences of individuals for conservation or utilization of these commodities (Bateman and Turner, 1993, p. 120). In order to arrive at an aggregate measure (total economic value), economists distinguish user value and non-user values (Pearce and Turner, 1990, pp. 129–137). The user value can be divided into actual user value and option use value. Recreation value and experience value belong to actual user value (Bateman and Turner, 1993, p. 121).

More complex are values expressed through options to use the environmental asset in the future. They are essentially expressions of preferences (willingness to pay (WTP)) for conservation of wildlife and landscape against some probability that the individual will make use of them at a later date. A related form of value is bequest value, a WTP to preserve wildlife and landscape for the benefits of the next generation(s). It is not a use value for the current individual valuer, but a potential future use or non-use value for his/her descendants (Bateman and Turner, 1993, pp. 121–122). Another form of non-use value is the vicarious value. It is the willingness to pay to preserve wildlife and landscape for the benefit of others (Pearce and Turner, 1990, p. 131).

Non-use values are more complex. They consist not only of bequest value and vicarious value, but also of the existence values. It is not very clear how they are best defined. According to Pearce and Turner (1990, p. 130), they are values which are taken to be entities that reflect people's preferences, but those values include concern for, sympathy with and respect for the rights or welfare of non-human beings and values which are unrelated to human use. These values are still anthropocentric but may include a recognition of the value of existence of certain species or whole ecosystems. Total economic value is, then, made up of actual use value, option value, bequest value and vicarious value plus existence value (Bateman and Turner, 1993, p. 122).

Including costs and benefits of changes in the availability of non-market goods and services in a cost–benefit analysis is difficult, but also very useful. However, there are only a few examples (Smith, 1992b, pp. 1076–1088; Hanley, 1995, pp. 39–55). It is possible to distinguish indirect and direct approaches to obtain information about the value of non-market goods and services. The indirect approach uses information on goods and services traded in markets to value the non-market goods or services under consideration. Examples of this approach are the travel-cost method and hedonic pricing method (Bateman and Turner, 1993, p. 123).

The direct approach uses surveys or questionnaires to elicit an individual's WTP for more of a public good or his/her willingness to accept (WTA) compensation, or compensation demanded, for less of a public good (for example, landscape and wildlife). This direct approach is a stated preference method

(Bateman and Turner, 1993, p. 123). The dominant non-market valuation method based on stated preferences is the contingent valuation method (CVM) (Mitchell and Carson, 1989, p. 2). According to Perman *et al.* (1996, p. 269), the contingent valuation technique consists of two forms, one employing an experimental approach, based on simulations or game analysis, the other using data derived from questionnaires or survey techniques. The contingent valuation technique provides the sum of use and non-use values, although the results depend on the questioning.

Table 9.5 gives an overview of the most important studies regarding valuation of agricultural landscapes and wildlife conservation. All the results are based on the CVM. The results presented and the literature partly included in Table 9.5 should be interpreted carefully. The context of the CVM studies is often quite different. Still, the following basic principles of valuation come up:

- Today's landscape is the preferred landscape (Willis and Garrod, 1993, p. 9).
- Most empirical studies reveal a decrease in WTP with increasing distance from a site (Willis and Benson, 1993, p. 275; Brouwer and Slangen, 1995, pp. 24–25).
- Landscape with a large variation receives a higher valuation. Traditional elements in the landscape are valued more highly than new elements (O'Riordan *et al.*, 1989, cited by Willis and Garrod, 1993, p. 2; Drake, 1992; Willis and Garrod, 1993). Areas with a high biodiversity value are not always valued most highly (Sijtsma and Strijker, 1995, p. 34).
- An accessible landscape is valued more highly than inaccessible areas of the same quality. Valuation has a strong anthropocentric element (Willis and Garrod, 1993).
- Areas that are perceived to be overcrowded, because of congestion, have a lower WTP (Willis and Garrod, 1993, p. 19).
- Besides the landscape value, the wildlife value is important to express for the valuation of diverse wildlife (Adger and Whitby, 1993; Willis and Garrod, 1993).
- Increasing landscape and/or wildlife area above a certain threshold reduces its per-hectare valuation.[6] A minimal size of wildlife conservation area, however, might still be important.
- Government wildlife areas have a lower WTP than private or club wildlife areas (Willis and Garrod, 1993, pp. 19–20).
- Expenditure on environmental goods is more income elastic than expenditure on food and housing. According to Willis and Garrod (1994, p. 216) and Hanley *et al.* (1997, p. 415), the income elasticity with respect to the amenity value of trees is 0.8, compared with an income elasticity of demand for housing of 0.6. Income elasticities of around 0.8 have been achieved for environmental goods by Komen *et al.* (1996) and Kriström

Table 9.5. Valuation of agricultural landscapes and wildlife conservation.

Authors	What has been measured	Aggregated WTP (Dfl year^{-1})
Beasley *et al.*, 1986	1. WTP to prevent moderate housing development from taking place on agricultural land	1. 402 ha^{-1}
	2. WTP to prevent large housing development from taking place on agricultural land	2. 824 ha^{-1}
Marinelli *et al.*, 1990	1. WTP to prevent a worsening of a natural park	1. 645 ha^{-1}
	2. WTA refraining from a visit to the park	2. 1614 ha^{-1}
Dillman and Bergstrom, 1991	WTP to prevent conversion of prime agricultural land to urban-industrial use (a quarter to the whole area)	60–154 ha^{-1}
Drake, 1992	WTP to prevent half of all agricultural land with:	
	1. Grain production	1. 203 ha^{-1}
	2. Grazing	2. 387 ha^{-1}
	3. Wooded pasture	3. 489 ha^{-1}
	from being cultivated with spruce	
Adger and Whitby, 1993	WTP	
	1. To retain the green belt in UK	1. 909 ha^{-1}
	2. To conserve wildlife in UK	2. 117 ha^{-1}
Spaninks, 1993	WTP to get a varied vegetation on ditch sides and an improvement of the situation of meadow birds	164 ha^{-1}
Willis and Garrod, 1993	WTP to preserve:	
	1. Today's landscape	1. 116.1 million
	2. Conserved landscape	2. 111.6 million
Willis and Benson, 1993	WTP for three nature reserves:	
	1. Derwent Ings	1. 1401 ha^{-1}
	2. Skipwith Common	2. 6366 ha^{-1}
	3. Upper Teesdale	3. 1223 ha^{-1}
Willis and Garrod, 1994	WTP to preserve South Downs Environmentally Sensitive Area (ESA):	
	1. User value	1. 135.3 million
	2. Non-use value (general public)	2. 86.6 million
Bonnieux and Rainelli, 1995	WTA compensation to change farming practices:	
	1. Decreasing intensification of dryland	1. 1115 ha^{-1}
	2. Decreasing stocking rate of cattle	2. 312 ha^{-1}
	3. Joining an organic farming network	3. 758 ha^{-1}
Brouwer and Slangen, 1995	WTP for the preservation of wildlife and landscape within the cultivated area in the Alblasserwaard	3630 ha^{-1}
Hasund, Chapter 6, this volume	WTP for preserving landscape elements of cultivated land in Sweden	330–445 ha^{-1}

Exchange rate 1994: 1 ECU = 2.15 Dfl.

and Riera (1996).[7] Expenditure elasticities for food are lower: 0.3–0.6 (Edgerton *et al.*, 1996). Moreover, income elasticities of agricultural products at farm level are lower than expenditure elasticities at consumer level.

This overview of the accessible literature gives only limited possibilities to value 'new' and 'improved' wildlife conservation area. The highest per-hectare value in Table 9.5 comes from the conservation of an agricultural area (Brouwer and Slangen, 1995). This implies that only 'reasonable' assumptions can be made. On the basis of all available information, we assume that 'new' wildlife conservation can be valued at 500 Dfl ha^{-1} $year^{-1}$. This is a point estimate (Fig. 9.2). Moreover, we assume that the increased values of existing wildlife conservation areas amount to 300 Dfl ha^{-1} $year^{-1}$. The existing wildlife conservation areas are about 150,000 ha (LEI/CBS, 1996, p. 20). All valuations are relative to a landscape where farming is the main activity and are assumed to include use and non-use values. We acknowledge that rough assumptions are made; therefore, some sensitivity analysis is used in the section about a model to calculate the consequences of the EMS under different assumptions. Moreover, we assume that it will take 10 years before the wildlife value is fully effective, with a linear increase during this period.

PRICE DEVELOPMENTS

An analysis of long-term developments should pay attention to price developments. Those developments are mostly driven by three main elements: the technological development in producing particular goods and/or services; the price developments of inputs used for production; and the demand for goods and/or services. Of course, these elements are not completely independent: a strong technological development with slowly changing demand puts pressure on the price of inputs. This is a development that is often experienced by agricultural production and holds, in particular, for labour, land and capital used in agriculture. To identify relative price developments, an analysis over a historical period provides insight into relative price developments. Table 9.6 provides this overview.

It is quite clear that price developments have both systematic and stochastic aspects. Systematic real-price decreases of agricultural product prices and current operating input prices are quite clear. Investments in machinery and land consolidation show small real-price decreases, land prices go up (but not always) and wages increase systematically. There are no indications of sudden price developments that differ very much from price changes over the last 10–20 years, unless the milk-quota system is abolished. On the basis of this analysis, we assume future price developments as indicated in Table 9.6. All these price changes incorporate a general inflation rate of 2% $year^{-1}$. Table 9.6

Table 9.6. Overview of real long-term price developments in The Netherlands (calculated on the basis of data of Central Bureau voor de Statistiek and Landbouw-economisch Institut).

	Annual real price change in % per year			
Variable	1966–1975	1976–1985	1986–1995	Total period
Agricultural products	−4.1	−2.2	−3.1	−3.0
Arable products	−3.8	−4.5	−0.6	−3.4
Milk	−3.6	−1.0	−2.2	−1.8
Current operating inputs	−3.0	−0.4	−1.6	−1.9
Feed	−3.6	−2.0	−3.7	−3.7
Wages	3.4	1.4	0.7	2.6
Investment in equipment	−2.2	0.1	−0.7	−0.8
Investment in land consolidation	−3.2	−0.0	−0.6	−0.8
Land price (arable)	0.8	2.2	1.21*	0.7
Land price (grass)	2.1	2.3	0.61*	1.7

*Period 1985–1994.

provides no information on non-market valuations of wildlife conservation areas and recreation. Here, an annual real-price increase of 0.5% year^{-1} is assumed.[8]

TAXES AND THE EFFECTS OF GOVERNMENT FINANCING

The EMS requires a large share of government financing, because of the quasi-collective good character of wildlife and landscape conservation. The government generates the required budget by taxing the economy. Taxing, however, is non-neutral in the economy (van Bergeijk *et al.*, 1993). Even if taxes are raised and returned to the consumers or taxpayers, the net national income might be reduced, because taxpayers adjust their behaviour in the market. In the international literature, this is characterized as the 'excess burden' or the 'marginal welfare costs of taxes'. Alston and Hurd (1990), Ballard *et al.* (1985) and Browning (1987) suggest excess burden percentages between 20 and 50. Here, we remain at the low end of this percentage: 20%.

Normally, taxes are directly related to generated income. If the net added value within an economy is reduced, because of decreased activities in agriculture and related agribusiness that are not fully compensated for in other sectors of the economy, this reduces the government budget. We assume a tax rate of 30% of the net value added generated by capital and labour.

A MODEL TO CALCULATE THE CONSEQUENCES OF THE ECOLOGICAL MAIN STRUCTURE UNDER DIFFERENT ASSUMPTIONS

It has been illustrated that many assumptions and uncertainties play a role in analysing the consequences of the EMS. In such circumstances, a model is very helpful to see the consequences of different information and assumptions.[9] Here, only some of its main characteristics are mentioned.

The model calculates the financial and economic consequences of the EMS. The analysis is limited to the area that will be purchased. No attention has been given to an area of about 100,000 ha (see, for example, Sijtsma and Strijker, 1995, p. 4) with management agreements between government and farmers.[10]

The model tries to incorporate all relevant items, even though some information is very uncertain. Several price developments have been included. We operate in nominal prices, based on the assumption of a 2% rate of inflation. The choice of price developments and also the discount rate is related to this inflation rate. A discount rate of 5% implies a 3% real discount rate. Annual price changes above 2% imply a real price increase.

A basic option has been defined according to the information and assumptions provided above. The purchasing period is 30 years and the total period of analysis is 50 years. All relevant information is included in Table 9.7.

The results of the base run, together with the results of some additional calculations, have been included in Table 9.8. The first column provides the base alternative, which indicates a financial burden of the EMS of 8.8 billion Dfl. The economic costs are only slightly lower: 8.0 billion Dfl. The government has not made a financial economic analysis of the EMS (MLNV, 1990, 1992). According to our calculations, the implications are not only a large financial loss, which is expected because of the creation of public goods, but also a very large economic loss. This illustrates the need for further analysis in this area.

The second column of Table 9.8 provides the results of the short-cut method or the free land-price approach, which has been explained earlier. Calculated economic costs are approximately 1 billion Dfl higher, using this method. This implies that a methodology that starts from the valuation of land in farming leads to higher economic costs. One of the reasons might be that the alternative revenue of labour and capital that leave agriculture has been overestimated; another reason could be the relatively high price of land because of tax incentives. The difference in financial burden between the two calculation methods is due to differences in taxes, which are included in the calculation. The short-cut method assumes no difference in taxes paid by existing and previous farmers.

The third column is based on the base run and the normal method. Because wildlife valuation is highly uncertain the valuations have been doubled,

Table 9.7. Assumptions used to calculate the consequences of the EMS: base alternative.

Description, dimension and name of variable	Level	% Change
Existing hectares of nature area (HECONG)	150,000	
Total number of hectares in the EMS (HEC)	150,000	
Period of transformation in years (PER)	30	
Period with increasing transformation (years) (PER1)	10	3.00
Period with constant transformation (years) (PER2)	10	
Period with decreasing transformation (years)	10*	5.00
Period between purchasing and converting land (years) (VNG)	3	
Period that converted land is at full nature value (years) (PERNW)	10	
Purchase price of land (Dfl ha^{-1}) (AP)	40,500	2.50
Financial revenue of land to be converted (Dfl ha^{-1}) (FOWGP)	400	0.50*
Net revenue of land to be converted (Dfl ha^{-1}) (NOWGP)	800	0.50*
Cost of converting agricultural land (Dfl ha^{-1}) (IP)	7,000	150
Maintenance costs (Dfl ha^{-1}) (OP)	1,500	2.00
Price increase of existing nature area (Dfl ha^{-1}) (PONG)	300	2.50*
Price increase of new area (Dfl ha^{-1}) (PNG)	500	2.50
Net value added of agriculture (Dfl ha^{-1}) (NTWLP)	3,750	1.00
Share of land in net value added (%) (AGNTWL)	30	
Use outside agriculture of labour and capital (%) (AL)	50	10†
Net value added of related production (Dfl ha^{-1}) (NTWOVP)	3,500	1.00
Alternative use of related labour and capital (%) (AOV)	75	5†
Reduced emission capacity in agriculture (Dfl ha^{-1}) (WECAP)	400	0.50
Costs of water boards (Dfl ha^{-1}) (WLAP)	150	2.00*
Discount rate (%)		5
Length of evaluation period (years) (PERT)		50
Tax percentage (%) (BELV)		30
Excess burden factor (FLOF)		1.2

*Implied value (no independent choice).
†Number of years before the percentage increases stepwise to 100.

Table 9.8. Overview of the total financial burden and the total costs (discounted and in million Dfl) of the EMS under different assumptions.

Variable	EMS base run	EMS base run (free land-price approach)	Higher valuation of nature
Total financial burden	8772	8521	8772
Total economic costs	8034	9004	6552

compared with the base run (1000 and 600 Dfl ha^{-1} $year^{-1}$, respectively). Although this makes the economic feasibility less negative, there is still an economic loss of the EMS, which equals more than 6.5 billion (10^9) Dfl.

DISCUSSION AND CONCLUSIONS

The negative benefits compared with the costs of the EMS are quite surprising if one realizes that a lot of effort has been put into this new policy. Also, the financial burden of the programme, which has been funded mainly by the 'green fund' and the regular budget of the Ministry of Agriculture, Nature Management and Fisheries (MLNV, 1994, pp. 106–121), is worrying, because a large part of the future financial burden should be provided by the provinces. The results differ also from those of Hampicke (1991), who came up with much more positive results for wildlife conservation.

Several aspects have been included in the analysis of the EMS in The Netherlands. Here, we followed a methodology where the most important costs and revenue items have been included. One often observes that researchers consider particular items as *pro memoria*, without valuation. Here, we operated differently by including several guesses of costs and benefits, but acknowledge that additional information might change the results substantially. Initial calculations of Slangen (1994a) and Oskam (1994) indicated quite high cost levels of the EMS. Later, Sijtsma and Strijker (1995) composed two reports on the EMS, one on the economics and one on the ecological and biodiversity aspects of the whole plan. In the economic analysis, they followed the methodology of Oskam (1994), but with different assumptions and without including the increased value of land under the EMS. They concluded that the costs of the EMS are much lower than the 8 billion Dfl (their results indicated approximately 4 billion Dfl) that have been calculated here. Given our present information, we see no reason to follow the numbers provided by Sijtsma and Strijker (1995).

A different approach would be to calculate the required valuation of wildlife conservation area to make costs and benefits of the plan over the total period of 50 years just equal to zero. Using the same assumptions, we calculated a valuation per hectare of 4835 Dfl. Measured in real prices, this implies a valuation of 6200 Dfl ha^{-1} $year^{-1}$ at the end of the 50-year period. This higher level is due to an annual real price increase of wildlife conservation area of 0.5%. Of course, the total calculation period might be important, because the benefits of the EMS are realized in the future.

Although several aspects have been included in our analysis, there are still a number of elements that require further attention. Most of them would increase the costs of implementing the EMS in The Netherlands.

1. The model opens the possibility of specifying a relationship between the land price used for the EMS and the area that will be purchased. This relationship has not yet been included in our analysis. Decisions on large areas will influence land prices. At present, most researchers base land prices on the situation with a low purchase level for wildlife conservation. This leads to an underestimation of the financial burden and the total costs of the EMS.

2. The decreasing value of landscape and wildlife in the remaining farm area have not been incorporated. This plays a role, because a part of agricultural production will be shifted to other areas and make land use more intensive. Moreover, one should realize that spatial planning of the EMS is mainly derived from ecologically coherent areas and from the need for green areas around cities. This aspect increases the costs of the EMS, compared with our calculations. It was argued earlier that a rent-bid function approach may lead to a different location of the EMS. From a dynamic perspective, the location of the EMS makes future infrastructure projects (roads, railways, town or city development) more inflexible.

3. The speed of establishing the EMS influences its level of cost. After an initial planning period of 30 years, the Dutch parliament requested the government to speed up the total planning period to 25 years (MLNV, 1992). Even if we assume that purchasing prices and converting prices of land remain at the same level, total costs of the plan would increase by 0.4 billion Dfl. It is likely, however, that both prices will increase due to an increased speed. Along similar lines, one might argue that alternative revenues of labour and capital in agriculture and agribusiness are lower with a higher speed of transformation.

4. The analysis considers a period of 50 years (or shorter). Wildlife development and conservation, however, require a long period. The influence of long-term developments depends on the discount rate. Moreover, it depends very much on the prices of 'old' and 'new' wildlife conservation area. In the base alternative, calculated costs increase from 8 to 10 billion Dfl if an infinite period of analysis is used.

5. Transaction costs of converting agricultural land to wildlife conservation areas have been neglected. Preliminary calculations of A. Van den Brink (1996, personal communication) for land consolidation projects and analyses of Whitby *et al.* (Chapter 8, this volume) in relation to Environmentally Sensitive Areas suggest large transaction costs: between 50 and 100% of the investment costs or operating costs of government projects. For The Netherlands, transaction costs for management agreements are 33% of the management compensation (Slangen, 1994b, p. 52).

6. We paid attention to the increased valuation of land in the EMS, but the values of houses within or near EMS locations may also increase.

7. Converting land to a wildlife conservation area influences agricultural production and therefore the budget costs of the EU. The Netherlands pays a (small) share of the budget costs and might therefore save some budget in supporting the EU funding in Brussels.

All these arguments (except the last two) lead to an underestimation of the costs of the EMS. They have not been considered by Sijtsma and Strijker (1995) and often require more information.

We conclude that the EMS of The Netherlands puts a high financial burden on the state budget and/or regional budgets. According to present information, the EMS gives large negative returns to Dutch society, unless extremely high valuations of wildlife conservation and recreation areas are included.

NOTES

1 The authors thank Stephan Dabbert, Alex Dubgaard and Martin Whitby for their stimulating comments on an earlier version, without making them responsible for any error or assumption.

2 Sijtsma and Strijker (1995, p. 6) use a total size of 144,000 ha. Moreover, they only distinguish nature development and nature preservation areas, but neglect the Strategic Green Projects, which have quite different costs and benefits.

3 Sijtsma and Strijker (1995, p. 59) calculate an average price of 3947 Dfl ha^{-1} year^{-1}, but they only consider nature development and nature preservation areas. Grontmij (1987) indicated a price of 82,000 Dfl ha^{-1} year^{-1} for recreation projects, which are included in the Strategic Green Projects.

4 Sijtsma and Strijker (1995, p. 59) assume and calculate annual maintenance costs of 800 Dfl ha^{-1}. Also, here, the recreation projects are not included in the analysis and their maintenance cost is estimated at *c.* 7000 Dfl ha^{-1} year^{-1}. Moreover, Loman (1996, pp. 34, 35) concludes that they underestimate the maintenance costs.

5 The price of the phosphate quota is between 20 and 40 Dfl kg^{-1}. It depends very much on the assumption of the length of the period that the phosphate quota will be in operation.

6 This element plays an important role in the analysis of Dillman and Bergstrom (1991) for the agricultural landscape in North Carolina. Colson and Stenger-Letheux (1996) interpret this as follows: 'quantity is not a determinant variable in environmental goods'. This observation is mainly related to the measurement method.

7 The main interest here is the income elasticity of demand for amenities with increasing real incomes in 'rich' countries. This differs from income elasticities based on a cross-section of people with different incomes (which mainly reflect the subsidy provided to different income groups of government-financed projects). Moreover, the 'good' to be valued is only partly recreation and consists of much more expenditure, such as walking, camping near 'nature parks' and so on, and also the non-use value of natural amenities. It is difficult to find relevant results for such items. According to Gratton and Taylor (1990, p. 48), activities like walking, swimming and fishing show (cross-section) income elasticities less than 1.

8 Besides such a partly empirical-based projection, we could also start with expenditure elasticities and assumptions about the availability of nature preservation areas. If we assume an expenditure elasticity of 0.8 for amenities, a population growth of 0.4% year^{-1} and an annual real growth of income per capita of 2% year^{-1}, this implies an increase of demand of 2% year^{-1}. Supply of amenities, however, will also increase because of the EMS. Depending on the reference quantity (and the loss of area for housing

and infrastructure), this will be around 1–2% $year^{-1}$. A guess of 0.5% real price increase of the nature preservation area belongs to a 'reasonable range'.

9 A detailed model description is available upon request.

10 Such an analysis could be made quite easily with the same model.

10

Landscape Values in Farming and Forestry Environmental Accounting (Area Scale Versus Enterprise Approach)

Edi Defrancesco and Maurizio Merlo[1]

INTRODUCTION: FROM NATIONAL TO ENTERPRISE ENVIRONMENTAL ACCOUNTING

To date, environmental accounting has been mainly applied in national accounts. It is acknowledged that an environmentally adjusted net domestic product would better identify true income, capture environmental services and account for damage and the depreciation of both man-made and natural capital stock, excluding defensive environmental expenditures (Lutz, 1993). Accepting the concept of sustainable development, Peskin and Lutz (1990), followed by Sammarco (1993), classify four approaches to national environmental accounting, with reference to the experiences and ongoing development of various countries (Peskin and Lutz, 1990):

- Identification and reclassification of environmental expenditure (Leipert and Simonis, 1989).
- Physical resource accounting or *approche patrimonial* (INSEE, 1986; Weber, 1986; Archambault and Bernard, 1988; Peskin and Lutz, 1990).
- Depreciation of marketed natural resources (Repetto *et al.*, 1989).
- Full environmental and natural resource accounts (Hueting, 1989; United Nations, 1993).

This chapter proposes a framework for farming and forestry enterprise accounting capable of including landscape and environmental values. Several options are advanced which represent progressive steps of sequential accounts:

- Identification of direct landscape or environmental expenditures.

- Making explicit hidden environmental values.
- Addition of off-site and non-market values estimated through consumer-surplus measures.

The above steps of environmental accounting must be reflected in both the balance sheet and the profit-or-loss account. The most advanced steps need 'satellite' accounts to be integrated as addenda into the traditional accounting systems.

The proposed accounting scheme, previously applied to individual farms and forest enterprises, has been enlarged to a system of rural enterprises located within a regional park – a multipurpose public forest, dairy farms, agritourism and recreational activities, including a golf course. This consolidated area approach allows comprehensive accounting capable of showing the flow of costs and benefits among different activities within the regional park – for example, landscape benefits created by farming and forestry resulting in revenue for tourism and recreation activities.

ENTERPRISE ENVIRONMENTAL ACCOUNTING

The issue of including landscape and environmental values in individual enterprise accounting has only arisen recently. Previously it was thought that non-market public 'goods' and 'bads', being part of social welfare, should be considered, at most, in national accounts at a macro level, rather than in individual enterprises. However, the growing awareness that environmental problems should be tackled at the core (individual production and consumption sites) has encouraged developments at enterprise or area scale. The market itself (green consumerism) appears to be pushing in this direction: take, for instance, the demand for goods produced in an environmentally friendly way and incorporating, at least in terms of image, landscape and environmental qualities. The acknowledgement of landscape values is supported by legislation on *appellation d'origine*, well established in Latin countries and now extended to the whole European Union (EU) by Council Regulations (European Economic Community (EEC)) 2081/92 and 2082/92 on product origins, quality and specificity. European Union Regulations 880/92 for eco-labels and 1836/93 for eco-auditing must also be mentioned. Recent proposals for stewardship certification in forestry and agriculture could also play a role. In addition, there is a debate on introducing tax declaration schemes capable of identifying the environmental impact of individual economic activities. Environmental auditing (Stern, 1993) is now considered an essential condition for all enterprises aiming to maintain the confidence of shareholders, to appeal to potential buyers and to merging partners (Bossi, 1992), to satisfy financial institutions and insurance companies (Spasiano, 1992) and finally to answer the expectations of the customers and the general public. Incidentally, competitive advantages seem to arise for economic systems and enterprises adopting environmentally friendly codes of practice (Porter, 1991). Again, it has been found in land appraisals that properties located in attractive landscapes have higher values.

This movement is creating a growing need for information. Enterprise environmental accounting could make a substantial contribution. Developments are, however, tentative (Larini, 1995). Certainly, it is remarkable how enterprises are moving from a passive, if not negative, attitude, where the environment is perceived as an external constraint, to a positive attitude, where the environment is seen as a new opportunity (Owen, 1992; Musu and Siniscalco, 1993). A growing number of enterprises try to anticipate institutional and social changes, moving from restoration or defensive policies to prevention (Dente and Ranci, 1992), if not ambitious strategies aimed at the complete management and control of the environment, adopting the sustainable-development philosophy to achieve the best balance between economic growth and conservation (Marangoni, 1994).

Methodologies and Schemes

Methodologies and schemes for enterprise environmental accounting are far from being well tried and tested, let alone codified and unanimously accepted. Experience up to now is mainly based on voluntary adhesion and, in any case, reflects difficulties in quantification and monetization of environmental impacts of individual enterprises. There is also clearly a search for a compromise between the need to inform the external world (company environmental reports) and the usual confidential nature of management accounting. The state of the art is such that even a common terminology is still lacking. On the other hand, there is a strongly felt need to make the various proposed schemes homogeneous, for the sake of transparency and comparability among enterprises and consistency with national environmental accounting.

Structures – objectives and contents – of enterprise environmental accounting are outlined in Fig. 10.1, which shows two main approaches:

1. The first refers to the so-called ecobalances, where the use of natural resources is described. Physical accounts and input–output matrixes are used. The results are sometimes shown by environmental-performance indicators (Bartolomeo *et al.*, 1995; FEEM, 1995), obtained by normalizing physical information according to economic variables (such as revenue, value added) or to the environmental maximum carrying capacity, which is, of course, very difficult to define (Marangoni, 1994). Sometimes environmental expenditure is also indicated (Musu and Siniscalco, 1993).

2. The second approach is derived from conventional financial accounting and tries to express environmental values in monetary terms. This requires a

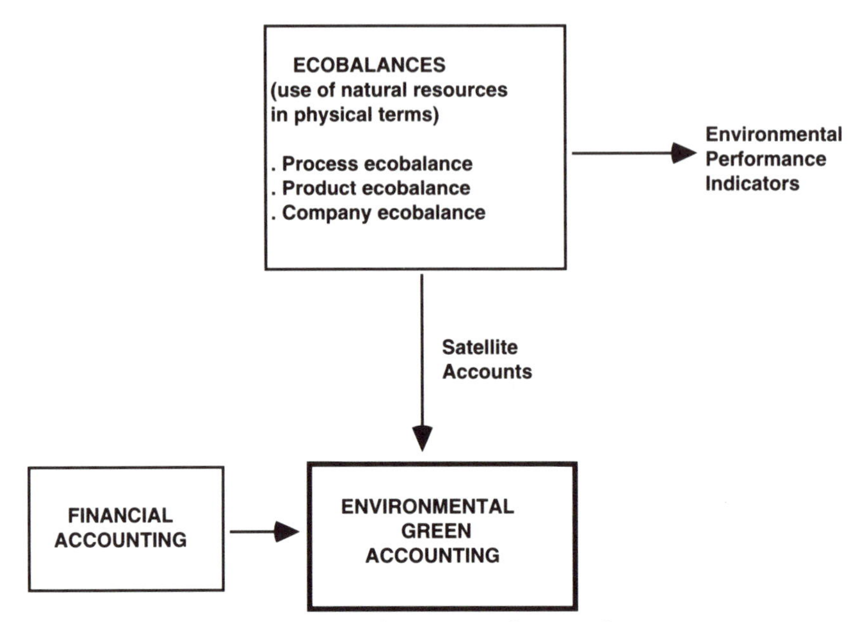

Fig. 10.1. Enterprise environmental accounting framework.

full integration of the environmental values into the accounting system. The profit-or-loss account should therefore show the so-called 'green' result, that is, an environmentally adjusted result. Possible procedures are, however, poorly defined, and sometimes the integration is achieved employing separate accounts – satellite accounts and/or addenda to the main accounting schemes – where environmental values can be shown (De Backer, 1992). However, the links between 'ecobalances' and 'environmental green accounting', as shown in Fig. 10.1, which should be given by satellite accounts, remain uncertain and far from being widely accepted and employed.

The pros and cons of the two main options (ecobalances and environmental accounting) outlined in Fig. 10.1 are well known. The various possible ecobalances present problems in quantification of environmental values, which, moreover, remain confined to physical values. It has therefore been argued that resource accounting in physical terms simply avoids the complexity of transforming resources into monetary terms (Dasgupta *et al.*, 1995). Meanwhile, environmental accounting applied rigidly can produce results that are not always credible. First, there are serious problems in monetization. Second, joint costs and revenues are difficult to allocate to 'environmental' cost or profit centres, even within sophisticated management account systems. On the other hand, integration of physical and financial values represents the only possible way to assess what level of profitability is environmentally sustainable and the real value of an enterprise's assets. Financial and economic analytical principles and methods are, however, flexible enough to allow consideration of environmental values, according to various degrees of certitude and the specific needs of management and public control. A 'progressive' integration of ecobalances – using satellite accounts – into the financial accounts is part of this process, which can create, gradually, a consistent environmental accounting. More specifically, traditional net profit or loss can be transformed into environmentally adjusted net profit or loss. This process could be conducted stepwise, initially including identification of environmental costs, then extended to resource accounting and to imputed costs, and finally considering off-site and non-market costs and benefits. It is rather interesting to note that the development from traditional to environmental green accounting is following the same path already pursued in national accounts, with regard to the whole economy, and previously by cost–benefit analysis (CBA), with regard to investment analysis. In fact, systems of national accounting (SNAs) have been developed into systems of integrated environmental and economic accounting (SEEAs), and CBA, initially based on market prices, has been 'extended' to include non-market environmental values.

A recent proposal by Siniscalco (1995) underlines the striking similarities of environmental accounting undertaken at national and enterprise scale. Bearing in mind the distinction outlined in the introduction at national level, the following four approaches to enterprise environmental accounting, ranging from 'light' to 'dark green', according to Gray's definition (Gray, 1993), can be mentioned:

1. *Identification and reclassification of environmental landscape expenditures.* This is already present in traditional accounting. Separate records are kept of expenditures undertaken for improvements (restoration and mitigation, stewardship, conservation and so on), paying environmental fines and taxes, meeting standards or coping with conflicts promoted by green movements. This approach – strictly based on market values – only calls for adjustment of existing accounting systems in line with a 'light green' option. Of course, the approach is helped by the creation of specific cost centres, where each type of environmental expenditure must be grouped. Allocation criteria for joint costs have to be defined, bearing in mind that several environmental services are produced jointly with marketable goods and services.
2. *Making explicit hidden environmental liabilities or assets.* This includes items such as future expenses for restoration, to meet more stringent legislation or to prevent future damage, as well as the imputed costs necessary to provide environmental goods and services. This approach highlights possible losses (depletion and/or degradation), but also gains (such as increased landscape beauty due to afforestation), which have monetary value. The approach can require some use of satellite accounts and/or addenda to traditional bookkeeping and year-end accounts. Of course, rapid and obvious depletion and degradation of natural assets (such as forest fires or dramatic cases of pollution), being

immediately expressed by market values, are generally included in traditional accounts. Meanwhile, less evident non-market depletion (such as the gradual lowering of the water-table) and degradation of natural assets (such as water pollution) or improvement of natural assets (such as a fallow year, green-manure production or increase of forest-growing stock) should be included in satellite accounts.

3. *Addition of non-market values.* This includes items such as natural stock depletion or degradation shown by forest inventories, extended to quality of tree stands, biodiversity, landscape impact and so on. In this way, elements of the total economic value (TEV) of environmental goods and services are taken into account. Measures can be referred to variation of consumer surplus resulting from valuation techniques, such as travel cost (TCM) and contingent valuation (CV).[2] Satellite accounts become essential. The physical and/or monetary information provided by satellite accounts is attached to the accounting system mainly through addenda, which supplement traditional accounting without altering its structure. It is interesting to note that satellite accounts and the various possible addenda can, to a certain extent, be assimilated into the memorandum accounts foreseen by Directive 78/660/EEC, including risks, obligations and warrants affecting the enterprise assets. Memorandum accounts are therefore supplementary to balance sheets and should highlight items that do not affect the traditional profit-or-loss account, but they are important for the overall impact and image of the firm. This approach, though highly innovative, has the inherent advantage of ensuring continuity of traditional accounting and the management information system. However, it must be clear that managerial objectives are extended to include a public view of landscape-based enterprises.

4. *Full integration of environmental and financial data.* This is where an integrated information system, aimed at completing environmental valuation of the enterprise, is set up according to a 'strong' sustainability view (Victor, 1991). In practice, this calls for complete readjustment of accounting procedures and is the most radical, or 'dark green', approach. However, approach 3 can lead to the same results, whenever account is taken of natural capital depletion or degradation, along with environmental damage, as well as benefits. Therefore, this fourth approach has rarely, if ever, been applied in its full extent and implications.

ENVIRONMENTAL ACCOUNTING: FROM INDIVIDUAL ENTERPRISES TO THE LANDSCAPE-BASED BUSINESS SYSTEM

Application of environmental accounting has been attempted in individual forest enterprises and dairy farms (Defrancesco and Merlo, 1996; Merlo, 1996) located in mountain areas of outstanding natural beauty, mainly regional parks. With specific reference to forestry, it has been shown that the above stepwise scheme, once adapted, can be a useful instrument for various managerial and policy purposes.

The conventional accounting results of the above-quoted application show the heavy financial losses of multipurpose public forestry – recreation, conservation and timber production. However, the first approach to environmental accounting, based on identification and reclassification, highlights that financial losses are caused mainly by recreation and conservation activities – real expenditures which cannot be met by adequate revenue. Timber production alone is generally capable of achieving a balance between costs and revenues.

The second approach referred to – making explicit hidden environmental liabilities or assets – shows environmental values – positive and/or negative – neglected by traditional accounting. These are, for example, gains associated with the increased volume of growing stock, and liabilities from past afforestation, which created artificial stands susceptible to collapse in case of natural hazards and pest attacks.

It is with the third approach to environmental accounting, based on inclusion of non-market and/or off-site values, making use of the TEV concept, that the overall economic picture of multipurpose forestry is fully represented and changed, highlighting its high social profit. In fact, recreation and conservation benefits are explicitly shown, thereby modifying the overall economic picture of forestry. Unfortunately, non-monetary benefits can only be measured through TCM and CV methods, which generate consumer surplus values that are not conventionally recorded in national income accounts.

The enterprise approach to environmental accounting, however, fails to show some possible internalization of public benefits attached to multipurpose forestry and agriculture. The case of recreational benefits 'captured' by other enterprises, especially tourist- and sport-related businesses, strictly based on agriculture and forestry so-called non-market benefits, is highly significant. Also, revenue linked to landscape quality and the local environment are not explicitly shown, although they are accounted for as revenue of traditional market products, as can be the case with *appellation d'origine* agricultural products. It has been demonstrated that their market value allows conservation of unique landscape, as is often the case with mountainous Less Favoured Areas, where agriculture can survive only due to quality products (Ferro *et al.*, 1995).

The problem can, however, be solved, as attempted in this chapter, through environmental accounting extended to all landscape-based businesses located in a certain area, a sort of enterprise group accounting, able to highlight the flow of public benefits and/or costs that go beyond individual enterprises. In other words, externalities (off-site and/or non-market effects), particularly those owning a use value, can be internalized at area scale. This area approach should, at the same time, avoid double counting and should stress the real contribution of the environment and the landscape to the income of the area.

THE CASE OF CANSIGLIO REGIONAL PARK

To illustrate environmental accounting enlarged to area scale, such as the system of landscape-based local businesses, an application is reported with reference to the Cansiglio Regional Park, located in Veneto Region, north-eastern Italian Alps. The enterprise system accounted for comprises the following units:

1. The Regional Forest Enterprise, producing timber, marketable recreation services, such as mushroom-picking permits, guided visits, 'green weeks', rent from tourism buildings and non-market services, such as unpaid recreation and environmental protection.
2. Four dairy farms producing organic milk according to techniques approved by National Organic Producers Associations, while fodder from meadows and pastures is produced according to Council Regulation (EEC) 2092/91 of 24 June 1991 on organic farming.
3. A milk-processing cooperative producing organic and *appellation d'origine* cheese, sold both on the wholesale market outside the park area and directly to consumers inside the park – 21% of the total.
4. Tourism activities, including two hotels, two restaurants (one of which is run by the milk cooperative) and two 'agritourism' enterprises run by two of the above farms.
5. Sports activities, including mountain bike hire (run from one of the agri-tourism farms), a golf club and an environmentally friendly ski centre based on cross-country ski tracks and alpine skiing, without artificial snow or snow chemical treatment.

The above activities can be easily viewed as an 'enterprise group'. This character is stressed by the fact that 80% of the land is in public ownership (Region), local farmers own cooperative shares and, last but not least, the enterprises in the park area are strongly dependent on income from tourism activities, to the extent of almost 400,000 visits per year.

The values shown in the sequence of 1995 year-end balance sheets and profit-or-loss accounts (Tables 10.1–10.4) have, as a starting-point, the existing enterprise accounting systems, reclassified following the mandatory layout prescribed to private enterprises by EU Directive 78/660. The year-end enterprise accounts have been consolidated following EU Directive 83/349. Consolidation adjustments have been made in order to avoid double-counting errors.[3] In particular, Table 10.1 gives the traditional profit-or-loss account and balance sheet at enterprise level and the consolidated one. The following tables (Tables 10.2, 10.3 and 10.4) modify and enlarge the year-end account

Table 10.1. 1995 year-end accounts (million lire; 1 European currency unit (ECU) = 1900 lire).

	Forest	Cheese cooperative (one restaurant)	Four organic dairy farms (two agritour)	Three hotel/ restaurants and skiing centre	Golf club	Consolidated
Profit-or-loss account						
A. Revenue from operating activities	813	2556	1767	1780	361	6479
Sales	787	2540	1092	1406	330	5358
Variation in stocks of finished goods and in work in progress		–29	587	332		890
Other operating revenue	25	45	88	42	31	231
B. Expenditure from operating activities	1339	2576	1679	1642	338	6776
Raw materials and consumables	68	1847	430	548		2187
Services	550	124	207	205	17	1103
Rent		6	33	56	9	13
Wages, salaries, social-security costs	579	456	304	242	145	1727
Depreciation of fixed assets	26	50	85	169	100	429
Variation in stocks of raw materials		36	569	336		940
Allocation to risk provisions						
Allocation to other provisions				2		2
Other operating charges	116	56	51	85	67	375
Operating result (A – B)	–526	–20	88	138	23	–297
C. Financial result	–40	–27	–63	–113	–20	–263
Financial revenue	2	4	2	1		9
Financial costs	–42	–31	–65	–114	–20	–272
D. Value adjustments in respect of financial assets						

Table 10.1. *Continued*

	Forest	Cheese cooperative (one restaurant)	Four organic dairy farms (two agritour)	Three hotel/ restaurants and skiing centre	Golf club	Consolidated
E. Extraordinary profit or loss	30	48	−11	4	−1	70
Extraordinary income (growing-stock and estate value)	30	58	10	6		105
Extraordinary charges		−11	−21	−2	−1	−35
Profit or loss pretax or compensation (A − B + C + D + E)	−536	0	15	29	2	−490
Tax						
Compensation (grants, subsidies and incentives)	600		78			678
Profit or loss for the year	64	0	92	29	2	187
Balance sheet						
Assets	15,947	1642	2628	1933	510	22,466
A. Subscribed capital unpaid		0				0
B. Formation expenses						
C. Fixed assets	15,858	675	1331	1390	500	19,724
I Net intangible assets (improvements value)	135	228	879	73	500	1816
II Net tangible assets	15,722	371	390	1310		17,793
(1) High standing forest (3197 ha), meadows and pastures (550 ha), buildings	17,980		35	117	520	18,652
(2) Plant machinery and other equipment	162	539	785	2067	220	3773
Depreciation provisions of fixed assets	−2420	−168	−430	−873	−740	−4632
III Financial assets (shares in participating undertakings)		76	62	7		115

Table 10.1. *Continued*

	Forest	Cheese cooperative (one restaurant)	Four organic dairy farms (two agritour)	Three hotel/ restaurants and skiing centre	Golf club	Consolidated
D. Current assets	90	955	1283	510	10	2683
I Stock		303	635	332		1271
II Debtors	26	652	431	139	10	1093
III Current financial assets						
IV Cash at bank and in hand	64		217	39		319
E. Prepayments and accrued income	0	12	14	33	0	59
Liabilities	15,947	1642	2628	1933	510	22,466
A. Capital and reserves	13,890	221	1350	360	70	15,860
I Subscribed capital	13,826	89	1188	202	68	15,284
II–IV Reserves		131	76	139		345
Consolidation reserve						60
V Profit or loss brought forward			–7	–10		–17
VI Profit or loss for the year	64	0	92	29	2	187
B. Provision for liabilities and charges	0	0	3	0	0	3
C. Provisions for pensions and similar obligations	256	66	28	26	0	376
D. Creditors	1802	1352	1204	1507	440	6141
(a) Amounts payable within 1 year	416	1271	951	663	440	3576
(b) Amounts payable after more than 1 year	1386	82	254	844		2565
E. Accruals and deferred income	0	4	42	40	0	86

Table 10.2. Approach 1: revenues and costs from ordinary operating activity separated from recreation and environment (million lire).

1995 Profit-or-loss account*	Forest	Cheese cooperative (1 restaurant)	4 Organic dairy farms (2 agritour)	3 hotel/ restaurants and skiing centre	Golf club	Consolidated
A1. Revenue from operating ordinary activity	668	1762	1515	0	0	3240
B1. Expenditure from operating ordinary activity	680	2051	1437	0	0	3422
Operating result from ordinary activity (A1 – B1)	–12	–288	78	0	0	–182
A2. Revenue from recreation and environment (see details)	145	793	252	1780	361	3240
B2. Expenditure from recreation and environment (see details)	659	525	242	1642	338	3354
Operating result from recreation and environment (A2 – B2)	–514	268	10	138	23	–115
Operating result	–526	–20	88	138	23	–297
C. Financial result	–40	–27	–63	–113	–20	–263
D. Value adjustments in respect of financial assets	0	0	0	0	0	0
E. Extraordinary profit or loss	30	48	–11	4	–1	70
Profit or loss pretax or compensation (A1 – B1 + A2 – B2 + C + D + E)	–536	0	15	29	2	–490
Tax	0	0	0	0	0	0
Compensation	600	0	78	0		678
Profit or loss for the year	64	0	92	29	2	187

Table 10.2. *Continued*

	Forest
Details of forest enterprise	
Revenue from recreation and environment	145
Rent and concessions from tourism buildings	103
Mushroom-picking permits	17
Guided visits and green walks	26
Expenditure from recreation and environment	659
Wages, salaries, social-security costs	133
Depreciation	10
Raw materials and services	419
Other costs	97
Services and other cost expenditure allocation to recreation and environment	
Landscape maintenance or improvements	138
Roads and hydrological works maintenance or improvement	31
Tourism buildings maintenance or improvement	113
Protection forest maintenance or improvement	142
Production forest maintenance or improvement	91

*See Table 10.1 for balance sheet.

Table 10.3. Approach 2: hidden environmental values (marked by §§) (million lire).

1995 Profit-or-loss account	Forest	Cheese cooperative (one restaurant)	Four organic dairy farms (two agritour)	Three hotel/ restaurants and skiing centre	Golf club	Consolidated
A1. Revenue from operating ordinary activity	668	1762	1515	0	0	3240
B1. Expenditure from operating ordinary activity	680	2051	1437	0	0	3422
Operating result from ordinary activity (A1 – B1)	–12	–288	78	0	0	–182
A2. Revenue from recreation and environment	145	793	252	1780	361	3240
B2. Expenditure from recreation and environment	659	525	242	1642	338	3354
Operating result from recreation and environment (A2 – B2)	–514	268	10	138	23	–115
Operating result	–526	–20	88	138	23	–297
C. Financial result	–40	–27	–63	–113	–20	–263
D. Value adjustments in respect of financial assets	0	0	0	0	0	0
E. Extraordinary profit or loss	30	48	–11	4	–1	70
Profit or loss pretax or compensation	–536	0	15	29	2	–490
Tax	0	0	0	0	0	0
Compensation	600	0	78	0		678
Profit or loss for the year	64	0	92	29	2	187
§§*Hidden environmental costs*	9					9
Risk of artificial stands depreciation	9					9
§§*Hidden environmental income*	70	123			75	145
Non-marketable growing-stock increase	70					70
Hidden landscape revenue from Cansiglio cheese		123				0
Hidden revenue from golf					75	75
§§Profit or loss adjusted for hidden environmental values	125	123	92	29	77	323

Table 10.3. *Continued*

1995 Balance sheet	Forest	Cheese cooperative (one restaurant)	Four organic dairy farms (two agritour)	Three hotel/ restaurants and skiing centre	Golf club	Consolidated
Assets						
A. Subscribed capital unpaid						0
B. Formation expenses						
C. Fixed assets						19,724
of which:						
§§ Landscape value from Cansiglio cheese						6
§§C. Hidden fixed assets						145
§§ Golf fixed assets value adjustment from 'normal' depreciation						75
§§ Non-marketable growing-stock increase						70
D. Current assets						2683
E. Prepayments and accrued income						59
Total assets (sum A–E and §§C)						22,611
Liabilities						
A. Capital and reserves						15,825
of which:						
§§ Net profit or loss adjusted for hidden environmental values						323
B. Provision for liabilities and charges						3
§§ Provision for risk in artificial stands						180
C. Provision for pensions and similar obligations						376
D. Creditors						6141
E. Accruals and deferred income						86
Total liabilities (sum A–E and §§B)						22,611

Table 10.4. Approach 3: linkage of financial accounting to environmental values through addenda (million lire).

1. 1995 Balance sheet (consolidated)

Assets		Liabilities	
A. Subscribed capital unpaid (forest public comp. unpaid)	0	A. Capital and reserves	15,825
B. Formation expenses		of which:	
C. Fixed assets	19,724	§§ Net profit or loss adjusted for hidden environmental values	323
of which:			
§§ Landscape value from Cansiglio Cheese	6	B. Provision for liabilities and charges	3
C. Hidden fixed assets	145	§§B. Provision for risk of artificial stands	180
§§ Non-marketable growing-stock increase	70	C. Provision for pensions and similar obligations	376
§§ Golf fixed assets value adjust from 'normal'	75	D. Creditors	6141
depreciation	2683	E. Accruals and deferred income	86
D. Current assets	59		
E. Prepayments and accrued income			
Total assets (sum A–E and C)	22,611	Total liabilities (sum A–E and B)	22,611

2. Addenda

Social assets		Social liabilities	
§§ Protection forest 160 ha	....	§§ Environmental debts toward society (protection and recreation forest, pastures, wildlife, biodiversity, game, etc.)	....
§§ Recreation forest and areas 800 ha	....		
§§ Wildlife (no. of species or quantity)	....		
§§ Game (no. of head)	....	§§ Net income or loss adjusted for non-market environmental benefits	....
§§ Biodiversity meadows and pastures	....		

3. 1995 Profit-or-loss account (consolidated)

A1. Revenue from operating ordinary activity	3240
B1. Expenditure from operating ordinary activity	3422
Operating result from ordinary activity	–182
A2. Revenue from recreation and environment	3240
B2. Expenditure from recreation and environment	3354
Operating result from recreation and environment	–115
Operating result	–297
Profit or loss pretax or compensation	–490
Compensation	678

Table 10.4. *Continued*

Profit or loss for the year	187
§§ Hidden environmental profit	136
§§ Profit or loss adjusted for hidden environmental values	323
Addenda: Social profit-or-loss account	
§§ Recreation	1329
§§ Mushroom-picking without permits	101
§§ Protection	320
less:	
Compensation	678
Non-market benefits net of compensation	1072
§§ Net profit or loss for the year adjusted for non-market environmental benefits	1396

according to the stepwise approach to environmental accounting previously described, ranging from 'light' to 'dark' green.

Conventional Accounts

The first profit-or-loss account and balance sheet (Table 10.1) refers simply to the enterprise year-end accounts. In the case of public forest, valuations have been made of fixed assets with reference to the real market: values, however, are rather conservative (almost half the market price), given the public-property status of all assets. Also, the farm year-end accounts are based on simplified accounting and prudent asset estimates. From the consolidated balance sheet, one can see that total assets (22 billion lire), mainly forest stands (11 billion lire) and buildings, meadows and pastures (5 billion lire), produce 6.5 billion lire revenue, mainly due to cheese and timber production and tourism-base services. The operating loss is equal to 297 million lire, resulting from the Forest Enterprise and the milk cooperative operating losses, which are not balanced by the positive results of dairy farming and tourism. The pretax result is negative (490 million lire), because financial costs are not fully compensated by the increased value of fixed assets – growing stock and estate value. It is interesting to note that the pretax loss is caused by the Forest Enterprise; however, it is covered by what are called compensations, that is, various contributions, coming mainly from the regional authority (600 million lire), justified on the ground that the forest provides a large amount of free public services contributing to people's welfare. Council Regulation (EEC) 2078/92 payments to organic farms (78 million lire) have been similarly added to the pretax results, improving the farmers' income. It must be noted, however, that the profit-or-loss account of Table 10.1, does not allow a clear distinction between the traditional activities (timber and dairy production) and the recreational environmental services. Nevertheless, these now typify the Cansiglio Park and its economy.

Identification and Reclassification of Environmental Landscape Expenditures

The profit-or-loss account (balance sheet unchanged) shown in Table 10.2 refers to the first approach to environmental accounting, where the traditional activity is separated from recreation and environmental functions. It should be noted that the approach does not require new values external to the existing accounting system, being merely based on a new aggregation of expenditures and revenue able to distinguish the two main different activities. This problem concerns mainly the Forest Enterprise, which does not fully distinguish the labour costs connected with environmental maintenance and the provision of recreational facilities from those directly connected to timber production. Labour costs have been allocated between ordinary activities and to recreation or environmental activities, adopting a relative contribution to the revenue criterion. Tourism activities of both farms and the milk-processing cooperative are accounted separately, so allocation criteria are needed only for overheads. This approach, however, requires, or at least is helped by, a revision of the accounting system toward management accounting (Jöbstl, 1995). As a result of this approach, it is possible to separate the heavy operating losses of the Forest Enterprise due to recreation and environmental activities (536 million lire) from timber production, which presents a positive result if the increased value of 'cuttable' growing stock is taken into account. The profit-or-loss account also shows that the negative result of recreational–environmental activities is balanced by public compensation. In fact, sales of environmental goods and services remain sporadic and scarce. The year-end accounts clearly indicate the need for new environmentally orientated marketing strategies.

The farms' positive operating results of ordinary activities are also separated from recreation. In the case of the milk-processing cooperative, the ordinary activity (milk processing) is negative (−288 million lire), due to the high price of milk paid to affiliated farmers. This loss, however, is balanced by the results of a shop–restaurant, selling traditional dairy products (268 million lire). The overall result, as demanded by the cooperative's statute, is therefore nil. This means that the operating result from the shop–restaurant has been included in the milk price. Remarkably, however, the price of milk paid to farmers is 38% higher than the market average. This milk-price differential, together with Council Regulation (EEC) 2078/92 aids, effectively contribute to farmers' income. Actually, they allow continuation of farming in the Cansiglio Park, assuring an average income of 18.5 million lire per farmer, almost comparable to non-agricultural income according to the EU definition (EU Directive 72/159 and Council Regulation (EEC) 797/85).

Making Explicit Hidden Environmental Liabilities or Assets

Table 10.3 shows hidden environmental values – specifically marked by §§. An environmentally adjusted profit is obtained equal to 323 million lire at group level. Various hidden environmental revenues have been considered (marked by §§), when already included in financial accounting:

1. The non-marketable growing-stock increase (prudently valued), which has been added to the forest balance sheet as hidden fixed assets.
2. Hidden landscape revenue from Cansiglio cheese in the milk cooperative profit-or-loss account (123 million lire). In fact, the *appellation d'origine* Cansiglio cheese is sold at higher prices than similar cheeses. The price differential can be attributed both to product quality and to consumers' willingness to pay for the product image, thereby internalizing landscape quality.[4] Furthermore, it may be noted that the present value of this 'hidden landscape revenue' (2% rate of discount) is equal to 6.2 billion lire. Therefore, a substantial part of the assets, as indicated in the financial balance sheet, could be attributed to 'active' dairy farming.
3. Again, 'hidden landscape revenue' can be found in the golf-club membership fees. In fact, the higher yearly fees (75 million lire) paid by members to balance an accelerated depreciation of fixed assets, may be considered as internalized recreation consumer surplus. Of course, the same value has been accounted in the balance sheet as a hidden fixed asset, being an adjustment of a 'normal' depreciation rate, with respect to the tenancy's life.

The above environmental values are all positive. Significantly, the negative adjustments are necessary to show the risks affecting existing artificial stands of spruce, which are liable to disease in plantation forestry, lacking biodiversity and natural protection. The artificial-stand risk, shown in the profit-or-loss account as a hidden annual depreciation,[5] should be covered by a specific provision to be incorporated as a liability in the balance sheet. This provision should be used in case of actual collapse of the artificial stands. However, the risk is also covered by annual management expenditures aimed at improvement of the artificial stands through planting of indigenous broad-leaved species, such as beech. The growing-stock increase imposed by management rules can also be seen as a concrete realization of a provision against this risk. It is interesting that this environmental accounting is supported by what may be seen as an old-established satellite account, namely, the forest inventory, which refers not only to stand quantity but also to quality.

Addition of Non-market Values

Table 10.3 takes into account enterprise assets or liabilities and products that have a market or can be easily expressed in market values. Meanwhile, Table 10.4 highlights non-market values, typically public goods. Therefore, the year-end accounts assume a full public connotation, open to welfare considerations. This means an enlargement of the balance sheet through shadowed addenda where natural resources received from society are indicated. These items are not quantified in monetary terms, but only physically, according to the logic of satellite accounts. Another addendum integrates the profit-or-loss account. It represents an adjusted (or enlarged) profit, able to account for the non-market and off-site flows of services – that is, the TEV. The operation seems acceptable to the extent that recreation and conservation add to social welfare; that is, they constitute a flow of utilities recorded in the same span covered by the profit-or-loss account. It should be noticed, however, that the net social profit is based on welfare estimation and derives from consumer-surplus measures – obtained from TCM and CVM. This flow of benefits is considered net of compensations, taken as an indemnity for the expenditures met to provide public goods and services.[6] This means that public compensations are omitted from the environmentally adjusted profit reported in Table 10.4 equal to around 1.4 billion lire, taken as an intrinsic component of the social-welfare function. Registration of this value in the profit-or-loss account of the Forest Enterprise, although as an addendum, may cause perplexities, at least as far as the amount is concerned. It seems, however, very important to include it, in order to show how the increased flow of services (annual welfare) contributes to social capital, which, in the end, constitutes human capital. Some authors tend to argue that only welfare variations matter (Linddal, 1995). On the opposite side, Adger and Whitby (1993) propose to modify the British agricultural product by adding the value of carbon fixation and non-market services flows, while deducting defensive expenditures. The arguments, therefore, are about consideration of total flows or limitation to variation. Apart from arguments having a certain rationale at nationwide level and for services provided by natural resources as such, it must be remarked that, in the specific case outlined in Table 10.4, inclusion of total yearly public benefits seems acceptable, because they correspond not only to resources but to environmental services that would be jeopardized by lack of landscape management – farming and forestry. Meanwhile, variation of stocks necessary to provide these services should be reported as asset variations in the balance sheets.

At area scale, non-market benefit from recreation (1.3 billion lire) is obtained by subtracting from non-market benefit derived from consumer-surplus measures (1.9 billion lire) the income already internalized by tourism and landscape-based enterprises (0.6 billion lire). In particular, non-market use values from recreation have been monitored since 1975, using TCM and CV methods, and quantified as 1.9 billion lire – on average, 5000 lire per visit. Prudently, the portion of these benefits captured by farms and tourism enterprises has been obtained, adding the operating results of tourism-based activities and the hidden recreation revenue of the golf-course to the hidden landscape revenue from Cansiglio cheese. The balance between estimated benefits from recreation and those already captured by the market is equal to 3400 lire per visit. This amount could be, at least partially, transformed into a market value – for instance, by measures, such as car parking tickets, that are able to capture consumer surplus without altering the existing property rights.

This valuation has been possible thanks to the area scale approach to environmental accounting. Its main advantage is certainly to show how landscape and recreational non-market benefits produced by certain enterprises have been, or could be, captured by others. In other words, the approach has highlighted how agriculture and forestry produce benefits that the consumers, at least partially, pay to other landscape-based enterprises – namely, the tourism activities.

FINAL COMMENTS AND CONCLUSIONS

A number of comments concerning economic, managerial and political issues may be drawn from the analysis and the tentative applications.

At the economic level:

- Environmental accounting allows internalization in the balance sheet of various near-market and non-market values connected with agriculture and forestry.
- Such internalization may be achieved stepwise, producing sequential balance sheets and profit-or-loss accounts, from 'light' to 'dark green' options.
- Options, however, can be separated, allowing clear demarcation between financial and environmental accounting: addenda to balance sheet and profit-or-loss accounts can be used to make clear differences, while satellite accounts can provide the necessary information.
- In any case, the various possible steps should reflect growing emphasis on environmental and social values connected with agriculture and forestry, making it possible to account for use, option and non-use values, the so-called TEV.
- In the most advanced steps of environmental accounting, assets borrowed from society and services provided to society can be highlighted within the logic of the most traditional financial accounting.
- With specific reference to the Cansiglio Regional Park, the consolidated approach allows the capture and shows in monetary terms flows of benefits that go beyond individual enterprises; an economic rationale is

therefore found for local compensation of environmental and landscape-based benefits.

At the management level:

- Functioning environmental accounting could allow for a comprehensive analysis of the various aspects inherent in multipurpose agriculture and forestry and the related management objectives.
- Financial ratios, cost-centres profitability and environmental indicators can be jointly developed, according to the same logic, allowing appropriate scrutiny of agriculture and forestry; also multicriterial optimization models can be better supported.
- Complete listing of agriculture and forestry outputs within the profit-or-loss account opens up an opportunity to explore potential markets for near-market or non-market outputs.
- The overall marketing strategy of the agriculture and forestry enterprises can therefore be re-examined, suggesting the best measures to achieve remuneration and/or compensation for all farming and forestry landscape-based benefits.
- Among such measures, appropriate means to capture consumer surpluses (environmental and landscape product development) can also be devised, in order to take full advantage of 'green consumerism'.
- The management of parks whose governing bodies consist of local authorities, amenity societies and various environmental pressure groups can be better informed on the conflicting objectives of different enterprises and actors, in order to find compromise solutions.

At the political and administrative level, environmental accounting can do the following:

- Improve public control of agriculture and forestry and help technical assistance to multipurpose landscape management.
- Facilitate the definition of compensation, grants and incentives, such as those foreseen by Council Regulations (EEC) 2078/92 and 2080/92; in addition, local compensation among the various enterprises can be better informed and regulated.
- Provide substantial support to the application of *appellation d'origine* and the various quality labels, including eco-label and stewardship certifications.
- Help define option and non-use values, suggesting regulations and incentives for those landscape-based services lacking any potential market and direct remuneration from the consumers.
- Contribute to better-informed definition and application of environmental policies, especially through the consolidation of the different year-end accounts of all enterprises and activities operating in a certain area or region. Compared with individual enterprise application, the present area-scale

approach has the main advantage of quantifying the environmental–recreational non-market benefits produced by an enterprise that has been, or could be, captured by another one. This quantification facilitates the definition of compensation among local enterprises and actors, according to the well-known 'subsidiarity principle', applied at its lowest possible level.

Finally, it should be noted that boundaries between financial and environmental accounting and between individual enterprise and area levels cannot be strictly defined. They overlap and have to be seen as continuous rather than discrete categories. Incidentally, some environmental issues have always been incorporated to a certain extent in financial accounting. This development from financial to environmental accounting can therefore be seen, and better accepted, as a step towards acknowledging the private or public status and objectives of different enterprises. Institutional issues and the state of property rights also contribute to defining boundaries and steps. The real-world development is such that environmental off-site and non-market outputs and even traditional inputs to agriculture and forestry, such as stewardship, are now becoming market outputs as far as payments are made, or can be made, by consumers and various other private, public or quasi-public bodies. Accounting, broadly speaking, has the task of registering, promoting and supporting this process.

NOTES

1. The authors wish to thank the different enterprises that have provided the data by allowing consultation of their bookkeeping. Of course, the present work must be considered as an illustration of environmental accounting, rather than an 'exact' operative approach. Elaboration of the basic data, including several estimates, is the responsibility of the authors.
2. Of course, it must be recognized that the CV method (CVM), at least theoretically, can estimate the whole TEV (use, option and existence values), while TC can only estimate use values.
3. Consolidated year-end accounts have been obtained in two steps: (i) totalling the enterprise year-end accounts; and (ii) adjusting the total to avoid overestimation of revenue, cost, assets and liabilities at consolidated level. Overestimation is due to double-counting of transfers of goods, services or money among enterprises of the same group. For example, milk produced by farms (706 million lire, included in farm revenue) and transformed by the cooperative (706 million lire included in cooperative processing costs) must be considered as an intermediate product at group level. Its value must be subtracted from the total of the consolidated revenue and consolidated expenses. Similarly, the infra-group rents (91 million lire) have been subtracted from both total cost of production and total revenue. Following the same path after summing up, enterprise balance sheets have been subtracted: (i) the cooperative amounts due to affiliated farmers for milk payments (164 million lire) and the same amount for (farmer)

receivables; and (ii) the farm's share of cooperative-subscribed capital from fixed financial assets (29 million lire) and the total subscribed capital unpaid (89 million lire). The positive difference between two such values (60 million lire) is accounted as consolidation reserve.

4. The *appellation d'origine* Cansiglio organic cheese is sold at 1300 lire kg^{-1} more than similar products (246 million lire in total). One-half of this differential can be assigned to environmental quality.

5. Artificial-stand risk depreciation covers 300 ha of artificial spruce stand, even-aged, 50 years old. Similar stands (200 ha) were attacked years ago by an insect (*Cephalcia harvensis*), making early felling necessary, with a loss per ha of 3 million lire, early cut and plantation of mixed forest. The probability of this occurrence within the next 20 years has been estimated at 20%, meaning that the risk depreciation amounts to 1%, so that the annual quota for 300 ha amounts to $3 \times 0.01 \times 300 = 9$ million lire. Although not applied, the risk depreciation can express the yearly risk insurance premium to existing stands due to the lack of biodiversity and natural protection.

6. This implies a double entry of compensations in the profit-and-loss account: firstly as a revenue increasing enterprise financial profit for the year and secondly as social expenditures to be deducted from the total value of recreational–environmental services (a sort of clearing transaction).

Private Provision of Public Environmental Goods: Policy Mechanisms for Agriculture

11

Robert D. Weaver

INTRODUCTION

Agriculture interacts directly with the environment, by utilizing environmental services as well as providing benefits in the form of environmental goods or services. Most often, these outputs of agriculture are public goods, for which consumption is non-exclusive and non-depleting, implying their benefits are realized jointly throughout society. An extensive literature has recognized this facet of agricultural production activities by considering agriculture's contribution to the conservation of prime agricultural land, wildlife habitat, surface and groundwater quality, atmospheric quality, recreation and a variety of related existence values. In many cases, these public goods result from agriculture not producing public bads or external costs, implying that their interpretation must be conditioned by an assignment of property rights. Here, this issue is left for another time and all such public effects as public goods are considered. The value of these agricultural public goods includes both market and non-market elements. None the less, these environmental public goods are largely the result of private decisions made by farmers concerning the management of their farm activities. This relationship among private production decisions and the provision of environmental public goods motivates a classic and strong case for public policy to ensure that an optimal level of environmental goods is provided. This chapter reviews the logic of this case for public policy, as well as the design or formulation of policy in the agricultural context.

The realm of public goods provided by agriculture is reviewed by Hodge (1991) and their dependence on private decisions was noted by Weaver and Harper (1993) and Weaver *et al.* (1996). This chapter proceeds directly to

 The Economics of Landscape and Wildlife Conservation (eds S. Dabbert, A. Dubgaard, L. Slangen and M. Whitby)

consider the case for public intervention and the complexity of the challenge of designing effective public policy. Here, of particular interest is the appropriateness of policy approaches that retain neutrality in their effects on farm outputs and input decisions. While such decoupling is of interest when income support is a goal, the nature of coupling between policy and production decisions is more complex in the case of support for private provision of public goods such as environmental services. Further, the policy design problem is complicated by heterogeneity across producers and asymmetry of information, as well as possible non-hedonic preferences held by producers. These conditions imply the regulator (agency) cannot observe their marginal costs of producing public goods or the marginal social benefits of provision of public goods. The third part of the chapter considers the design of policy under these conditions and offers guidelines for alternative approaches to managing the private provision of environmental goods by agriculture.

THE PUBLIC INTEREST IN ENVIRONMENTAL BENEFITS OF AGRICULTURE

The notion that private agricultural production activities generate public-type effects has a long history of recognition. European Community law responded to political support for public action to stimulate private provision of such public effects through the use of a variety of agri-environmental programmes. These programmes have been used as standardized instruments – voluntary agreements with producers to take particular actions that are presumed to lead to the achievement of particular environmental goals. Beaufoy (1994) and Bonnieux and Weaver (1996) provide useful reviews of these programmes. The focus of this section is to consider the public economics of the rationale for a public role in the provision of environmental benefits from agriculture.

Following Weaver (1996), we note that agriculture involves a production process which is joint in private and public inputs and outputs:[1]

$$G(Y^i, Q^i, E^i, Z, \theta^i) = 0 \tag{1}$$

where the superscript i indicates the variable is associated with the ith producer, Y^i is an $M \times 1$ vector of private good outputs, X^i is an $N \times 1$ vector of private-good variable inputs, E^i is an $L \times 1$ vector of environmental effort, θ^i is a $J \times 1$ vector of flows from quasi-fixed private factors of production, Z is a $K \times 1$ vector of public-good input flows, which include environmental services, and Q^i is a $J \times 1$ vector of public-good outputs. In the agricultural case, we interpret Z as including service flows from non-site-related resources, such as climatic and hydrological processes, and Q^i as environmental effects that are site-related in origin only. The function $G()$ is assumed to be continuously twice differentiable, with non-negative output gradients, non-positive input gradients, concave in outputs and convex in inputs. The joint-production function (eqn 1) is purely static; all variables are measured contemporaneously. Within this notation, it is clear that private decisions influence the extent of production of public goods. In the agricultural case and for the purposes of this chapter, Q^i is viewed as environmental goods, which include the full realm of market and non-market benefits associated with landscapes, biodiversity and habitat.

The private interest in private provision of these goods follows from a consideration of the agent's cognizance of public goods involved in production, as well as the agent's objectives (Weaver, 1996). Where the agent is myopic or is uninformed of the public-good interactions implicit in eqn (1), the agent might perceive a myopic joint-production function that involves only private goods:

$$F(Y^i, X^i; \theta^i) = 0 \qquad (2)$$

Between the extremes of eqns (1) and (2), alternatives might exist. As for private objectives, define privately appropriable profits from an optimal production plan (Y^* X^* E^*) as:

$$\pi^* = P'Y^* - R'X^* - q'E^* = \pi(P, R, q; Z, \theta^i) \qquad (3)$$

where P, R and q are vectors of unit prices, exogenous to the decision-maker.[2] In the most general case, agent objectives may involve both profits and preferences over other outcomes of production decisions. Where objectives are limited to profits, the agent can be viewed as a selfish hedonist, focusing only on the privately appropriable aspects of outcomes.

In contrast, Weaver (1996) considered a series of other possibilities, where the agent holds preferences over the public-good outputs associated with production. The most general of these cases is that of impure altruism, where preferences are held both over the agent's own public-good output Q^i and over that of other agents in a group of other agents. That is:

$$U = U(\pi, Q^w, Q^i; \Phi) \qquad (4)$$

where $Q^w = \sum_{j \in W} Q^i$ is the level of public-good output produced by agents in set W, which might include all individuals producing environmental effects or individuals in a social reference group, such as a watershed or an environmental or farm organization. The existence of these types of non-hedonistic preferences by producers has been implicitly acknowledged in numerous studies that have cited the role of goodwill of producers as a determinant of their decisions to participate in voluntary programmes; see, for example, Weaver (1996) for an overview of this literature.

Within this notation, we can now consider private provision of public goods by agriculture. As demonstrated by Weaver (1996), private environmental effort will be zero for myopic agents, and the level of private provision of the public goods will be determined by the myopically, privately optimal input allocation of X. That is, the jointness of production defined by eqn (1) will always imply public goods may be produced, given that input application is non-zero. These results follow intuitively from the fact that, in this case, the agent faces no incentives (price or productivity) to control Q^i. Now, suppose the agent is non-myopic and is cognizant of eqn (1), though remaining a selfish hedonist. Again, the agent faces no price incentives to produce the public good; however, to the extent that the private-good productivity of the agent's private inputs is increasing in the level of public-good outputs, the agent may find incentive for positive environmental effort and expand production of the public goods beyond that which myopic agents would provide. Finally, where the agent is either an altruist (Q affects preferences) or an egoistic hedonist (Q^i affects preferences), private provision of the public goods will further expand. However, this private provision of public environmental goods will always be set at a level that optimizes private interest, not necessarily social welfare.

Thus, at a theoretical level, the conclusion is warranted that private provision may lead to either under- or oversupply of the public environmental goods relative to socially optimal provision. Intuitively, the agent's preferences will play an important role in determining the nature of the final outcome. Green-activist farmers could be expected to oversupply the public goods, while others might set private environmental effort to zero and undersupply the public goods. To add precision, define the social-welfare function S as:

$$S = s\,[V(Q, Y) - P'Y] + (1 - s)U \tag{5}$$

where $Q = \Sigma Q^i$ is the aggregate supply of public environmental goods; $Y = \Sigma Y^i$, the aggregate supply of private goods; $\pi = \Sigma \pi^i$, the aggregate level of profits; $V = \Sigma V^i$ the aggregate consumer utility from consumption; s, the social-welfare weight for consumers; and aggregate producer welfare is $U = \Sigma U^i$. Where producers are hedonistic, $U^i = \pi^i$. We specify social welfare to be dependent on producer welfare rather than profits, supposing that society places identifiable weight $(1 - s)$ on producer welfare. Alternatively, a more traditional specification would view all social welfare as equally weighting consumers and producers, implying, for the hedonic case:

$$\tilde{S} = V(Q, Y) - P'Y + \pi = V(Q, Y) - R'X - q'E \tag{6}$$

The first best supply of public environmental goods and private environmental effort follows from maximizing social welfare (eqn 5), subject to eqn (1) and the specification of producer preferences. Optimal private provision follows from maximization of producer utility, subject to either eqn (1) or eqn (2), as appropriate. By comparison of the first-order conditions for these problems, the economic rationale for a public policy to reduce the discrepancy between privately and socially optimal levels of the environmental goods is clarified.

What remains is a consideration of alternative approaches. Before moving on, it is of interest to note that, while finding welfare-improving policy mechanisms is of interest, public debates over recent years raise two concerns of particular relevance. First, policy to alter private supply of public environmental goods may result in increased agricultural incomes, raising the question of whether income support is, in fact, the policy objective. Clearly, subsidies would increase incomes whenever they exceed the private marginal cost, net of any private marginal benefits associated with the change in private provision. However, within the context of the regulation literature, whenever asymmetric information characterizes the mechanism-design problem, agents will be able to collect rents based on their private information. A second issue of concern is the extent to which policy is decoupled from private-good allocation decisions. In the agricultural case, marginal taxes or subsidies designed to affect changes in private provision of environmental public-good supply distort production plans.

PUBLIC ECONOMICS OF MANAGEMENT AGREEMENTS

A variety of overviews of alternative mechanisms for managing the supply of environmental public goods from agriculture are available (see, for example, Hodge, 1991; Crabtree and Chalmers, 1994). Here, given our interest in considering policy that affects landscape- and biodiversity-related environmental goods, we focus our consideration on investment grants and direct payments for standards. Investment grants involve cost sharing by a public agency[3] for private costs incurred in making an investment that effects a desired change in the level of public-good provision by the agent. In the USA, the UK and France, this method has been extensively used to stimulate investments that alter or change technologies and result in increased environmental benefits (or reduced costs) or which implement particular structures, such as ponds or wildlife areas, that generate environmental benefits. Grants are awarded to producers who implement prescribed investments. In most cases, these grants are proportional to direct and estimated labour installation costs. The grants are often lump-sum, one-time direct payments that are conditionalized on completion of the investment implementation. Direct payments for standards involve annual payments for implementing production practices or private input or output mixes which meet a prescribed standard. In this sense, this approach constitutes a direct payment for accepting an input constraint, an output quota, or both. Importantly, this approach involves a voluntary acceptance or participation in the programme and uses a direct payment that does not vary across individual agents.

A third approach is to use individualized payments for individual effort, or management agreements. This last type of mechanism has a long history of use in private-property management by renters within the context of covenants on leaseholds or tenancies. Its application to management of the public interest in private lands was described by Hookway (1967) and Leonard (1982), among others. Over the past decade, its use has been extensive in the UK, where payments have been offered for voluntary compliance with long-term prescriptions for conservation practices (see, for example, Leonard, 1982; Curtis, 1983; Whittaker *et al.*, 1991). Whittaker *et al.* (1991) analyse these agreements as contracts to reduce output; however, the restrictions on land use involved in management agreements may be quite general, involving restrictions on several inputs or outputs, on timing of operations or on intensity of land use. In all cases, these restrictions are selected by the agency as a means of achieving particular conservation goals or, in terms of the notation of eqn (1), of achieving particular changes in Q^i. While in concept both prescriptions and payments would vary across farm settings, in practice in the UK prescriptions have been standardized within roughly homogeneous areas (for example, in Environmentally Sensitive Areas (ESAs)). In the remainder of this section, the microeconomics of these alternative approaches will be analysed and compared. In particular, their relative welfare implications (including social welfare and producer income), necessary compensation levels, efficacy in achieving changes in environmental goods and private-production plan distortion will be considered.

Investment Grants

The feasibility of direct public provision of environmental goods from agriculture has limited applicability, given the wide geographical scope of agricultural activities. None the less, direct purchase of particular types of environmental goods or the environmental characteristics associated with them (elements of Z in eqn (1)) has been pursued in both Europe and the USA (see, for example, Colman *et al.*, 1992). As has been noted in the US literature (Hodge, 1988), this approach may involve a transfer of property rights through a direct purchase by a public agency. The resulting rights are then provided as a public good to society in general. Examples include access to farmland for recreation, hunting or fishing, protection from urban development through transfer of development rights or preservation through transfer of rights of use.

A less direct approach to public policy to effect private provision of environmental goods involves the agency offering to finance changes in private quasi-fixed factor endowments through investment grants. Colman (1991) and Crabtree and Chalmers (1994) provide a description and examples of this approach. In this section, the objective is to place this approach within the context of the more general problem of public approaches to encouraging the private provision of public environmental goods by agriculture. In general, the approach involves the agency offering to finance (wholly or partially) investments that change elements in θ^i to enhance the environmental-good productivity of private production decisions. From this perspective, the approach may represent either standardized or individualized approaches. Under the standardized approach, a standard set of pairs, involving proposed changes in θ and fixed, direct payments T, is offered to agents. This approach is extensively used in the USA, within the context of investments for conservation structures (installation of stream-bank fencing or vegetative-filter strips) and investments (animal confinement area runoff structures), through the Agricultural Stabilization and Conservation Service of the US Department of Agriculture. Similar programmes are also offered to farmers through the US Forest Service, encouraging installation of enhancements to wildlife habitats such as ponds, wetland areas or particular vegetation. Under the individualized approach, agent-specific investments and payments are negotiated, providing a pair $(d\theta^i, T)$. Clearly, relative to the standardized approach, the individualized approach is substantially more expensive to implement, although net benefits may rationalize its use.

To understand the public economics of these approaches, consider the choice problem faced by the agency that seeks a second-best policy. For either approach, the agency faces asymmetric information. That is, while the agents have knowledge of θ^i, this information is not observable by the agency. Instead, we assume the agency has knowledge of the distribution of θ^i, $f(\theta)$, across agents. Thus, while the agency is assumed to have perfect knowledge of the technology (eqn 1) faced by agents, the specific nature of production possibilities faced by agents is conditional on θ^i which is assumed unobservable by the agency. For standardized investment grants, the agency would seek to identify a menu of pairs $(d\theta^i, T)$ that maximizes its perception of social welfare – for example, eqn (6). The agency would want to identify these pairs (contracts) such that they could be offered and agents would be allowed to choose the contract that satisfies their objectives. Given asymmetry in information held by the agency, to ensure that agents choose contracts that are optimal with respect to their current endowments θ^i, the agency would choose the menu of contracts to be offered such that: (i) acceptance of the contracts by agents would be rational (individual rationality); and (ii) agents would choose the contract that is optimal with respect to their true endowment level θ^i (incentive compatibility). This problem can be stated as the agency choosing $(d\theta^i, T)$ to solve:

$$\max S = E_\theta\{s[V(Q, Y) - PY] + (1 - s)\pi(Q, T; \theta)\} \tag{7}$$

subject to:

(a) incentive compatibility $\pi(Q, T; \theta, \theta) \geq \pi(Q, T; \theta, \theta^R)$

(b) individual rationality $\pi(Q, T; \theta) \geq \hat{\pi}(Q, T; \theta)$

where $\pi(Q, T; q) \equiv \max \pi = PY - RX - qE + T$

subject to $G(Y^i, Q^i, X^i, E^i, Z, \theta^i) = 0$, θ^R is the level of θ reported by the agent through selection of a contract pair, and $\hat{\pi}(Q, T; \theta)$ is the level of profits attainable if the agent does not participate in the programme.

In comparison, the approach of individualized investment grants follows when it is assumed the agency has limited knowledge of the technology (eqn 1), which varies across agents, even when conditionalized on θ^i. This might be the case when substantial variation in enterprise mix and environmental conditions (Z) occurs across the population of agents, implying they operate different technologies. For this case, the mechanism-design approach is not useful, since a single menu of contracts cannot be identified to satisfy such a diverse set of agents.

The specification in eqn (7) provides a useful basis for assessment of the usefulness of the investment-grants approach. Given the caveat just discussed, the approach appears attractive and tractable for cases where changes in private, quasi-fixed factor endowments θ^i offer potential for changes in the level of private provision of public environmental goods. That is, the efficacy of this approach relative to others depends ultimately on technical responses $dQ^i/d\theta^i$ and $\partial Q^i/\partial\theta^i$, which are interpretable as the environmental public-good total response and productivity of private quasi-fixed factors, respectively. This issue is considered within a different context by Weaver *et al.* (1996), who point out that this response may be quite small for private variable inputs. While the productivity may be non-zero, a change in θ^i can be expected to induce changes in the production plan, which might further reduce the response of Q^i as other private-good adjustments are made. Thus, it may occur that $dQ^i/d\theta^i < \partial Q^i/\partial\theta^i$. Within the context of land set-aside programmes to achieve supply control, this type of slippage is apparent, as farmers substitute other private inputs for land that is rationed by the policy.

Direct Payments for Standards

To begin analysis of this option, we must reconsider the agency's selection of the uniform prescription. From eqn (1), it is clear that agents are heterogeneous, given different endowments (Z^i, θ^i) and preferences. However, suppose the agency presumes hedonism, ignoring producer preferences for voluntary private provision of the environmental good. Suppose the agency derives some goal for the vector Q. The agency then faces the problem of choosing a menu of prescription and payment pairs. This problem is analogous to that considered under investment grants, such as eqn (7), although in this case the prescription represents constraints on private, variable input use and, possibly, quotas on private-good outputs.

To reconsider this problem in this slightly different context, we elaborate on the consideration given to investment grants. In practice, the agency faces two significant problems: heterogeneity across agent endowments of private, quasi-fixed factors and imperfect information concerning the production technology (eqn 1). Further, even in the face of homogeneous producers, for a given goal, say Q_m, eqn (1) offers no unique prescription for the private-good production plan. This is the essence of an important inefficiency involved in the use of agency-selected prescriptions. Supposing that a technically efficient private-good production plan can be identified, it will not be economically efficient for most producers and will impose substantial adjustment costs on those producers to switch to that private-good mix. These adjustment costs can be substantial and may very probably discourage participation in a voluntary programme. Graphically, the agency would be asking the producers not only to restrict particular private goods, but to shift to a specific production plan.

To further consider the option of direct payments for uniform standards, we step aside from this problem of agency choice of the prescription and suppose the agency offers the producer a prescription for a subvector of private inputs, X_m. Based on the agency's perception of technology (eqn 1), the agency could simply calculate a vector of marginal taxes t_m on the elements of X_m necessary to induce private production plans consistent with X_m. This approach would by definition achieve the prescription, but at a cost, which includes the technical inefficiency of X_m as well as the distortion of production choices by marginal taxes.

Alternatively, to avoid these inefficiencies, the agency could choose vectors of uniform quotas Y_m, standards X_m, and fixed payments T_m that would maximize social welfare. This design problem would take the following form:

$$\max \widetilde{S}^g = s[V(Q, Y) - PY] + (1 - s)\pi^g(Y_m, X_m, T_m) \qquad (7')$$

$$\text{where } \pi^g(Y_m, X_m, T_m) \equiv \max\pi = PY - RX - qE + T_m$$

$$\text{subject to } G(Y^i, Q^i, X^i, E^i, Z, \theta^g, Y_m, X_m) = 0$$

Here, the superscript g indicates an agency perception of the representative agent condition. Clearly, the solutions of this problem require the agency to set marginal social benefits equal to marginal social costs; however, it does so along the locus of production plans that the representative agent (as described by G conditioned by θ^g) would find privately optimal. Intuitively, the agency would be able to announce a pair, consisting of the production prescription (Y_m, X_m) and standard payment T_m, which it perceives will result in private provision of Q that solves eqn (7).

While appealing, this mechanism suffers from several flaws. First, to the extent that agents are heterogeneous, their actual profit functions will differ from that perceived by the agency. As a result, agents will fall into three classes: those who in the absence of the programme would choose (Y_m, X_m), those who will find the participation optimal but requiring restriction of their production plan and those who will find participation suboptimal. Clearly, it is possible to set the prescription too restrictively, given the payment, and no agent will participate. Alternatively, if these parameters are set too generously, all agents may participate and yet with production plans they might have implemented even in the absence of the programme. The increment of profit resulting from participation would represent an information rent. Intuitively, the direct payment provides an upward shift in the profit function for all production plans in compliance with the prescription, and the prescription will distort production plans from their privately optimal level. Thus, under compliance, the profit function will also be twisted and not be parallel to the non-compliance profit function. However, substantial rents will be paid to agents who comply, even though they would have operated production plans within compliance, even in the absence of the direct payment. These possibilities are apparent from the shift and twist of the profit function implied by the programme. Importantly, many agents that find it suboptimal to participate may have endowments, which imply substantial opportunities for affecting socially optimal changes in Q^i; however, the inefficient programme fails to catalyse these agents to participate, and social benefits are lost.

Management Agreements

Whittaker *et al.* (1991) provide a graphic analysis of the welfare implications of the private-output implications of management agreements. Their analysis does not consider the individualized nature of the agreements and assumes management-agreement covenants will restrict output to minimize the difference between private and social cost of production. Further, they presume that environmental public-good supply is proportional to private output levels. Social cost includes the external costs of negatively valued environmental outputs (bads), while private costs ignore such effects. The objective is focused on cost because of political interests of environmental groups, and output is restricted to Y_m, a level which is below the privately optimal supply (Y_p) and at which marginal social and marginal private costs intersect. In contrast, maximization of social welfare by setting marginal social cost equal to marginal social benefits would restrict output to a level (Y_s) below Y_m. They also note that price support would shift marginal private benefits upward, increasing the cost of the supply reduction. From their analysis, it is clear that, under price support, $Y_p > Y_m > Y_s$. However, if, in the absence of price support, marginal social benefits of production are equal to the competitive market price, these inequalities would be $Y_m > Y_s > Y_p$. That is, private decisions would lead to an undersupply of the private output. This analysis follows from supposing that social welfare is defined by $\widetilde{S}$ in eqn (6) and producers are profit-maximizers. Under these conditions, privately optimal environmental effort is set above zero only when it generates private marginal benefits that exceed its unit cost. Environmental-good outputs are set to ensure that their marginal productivity is zero ($\partial G^i/\partial Q^i$).

In contrast, where external effects are not proportional to private output, but result from a joint private/public-good production process, such as eqn (1), a true trade-off exists between private- and public-good production. Policy cannot be designed simply by consideration of private-output levels. Marginal social- and private-cost curves for private outputs would shift as private effort, input and other output levels change. Where social welfare is defined by eqn (5) and producers are not simply profit-maximizers, marginal benefits differ from the competitive price and include marginal benefits to producers, based on their preferences for producing the public environmental good. In this case, while producers would privately supply environmental public goods, the level of supply would not be Pareto-optimal. We conclude that such a mechanism is not second best.

As an alternative to management agreements that involve uniform private-good prescriptions (for example, as considered by Whittaker *et al.*, 1991), individual management agreements are considered next. To simplify without loss, we limit our focus to a rationing of individualized inputs, which are offered to the agent in return for individualized payments. (The problem is easily restated to include quotas on output.) We assume the agency has as an objective of controlling Q; however, we assume that environmental outputs are unobservable without great cost. We translate this problem into a mechanism-design problem as follows. At the root of the heterogeneity across agents lie differences in preferences and quasi-fixed factor endowments. To simplify, assume agents have similar preferences and differ only by a scalar factor endowment, θ^i. We assume that the technology (eqn 1), the environmental input flows in Z and the density function $f(\theta^i)$ describing the distribution of θ^i are observable by the agency. In contrast to the direct payment for standards approach, we assume the agency recognizes the heterogeneity across agents in the design of the programme. Suppose the agency offers a menu of prescription, payment pairs designed to maximize welfare subject to the constraints that it induces the agent to truthfully reveal θ^i (incentive compatibility), it leaves the agent at least as well off as without the programme (individual rationality), and it is consistent with utility or profit maximization. By use of a fixed, direct payment, the agency also ensures that private decisions will not be distorted by marginal incentives. The agency problem can then be stated as one of choosing a menu of contracts defined by the pair (X, T) of individualized prescriptions over private goods and direct payments that would maximize social welfare.

Consider firstly the case where the agents are pure hedonistic profit-maximizers, where the design problem would take the following form:

$$\max S = E_\theta\{s[V(Q, Y) - PY] + (1 - s)\pi(Q, X, T; \theta)\} \quad (8)$$

subject to:

(a) incentive compatibility $\pi(Q, X, T; \theta, \theta) \geq \pi(Q, X, T; \theta, \theta^R)$

(b) individual rationality $\pi((Q, X, T; \theta) \geq \hat{\pi}(Q, X, T; \theta)$

where $\pi(Q, X, T; \theta) \equiv \max \pi = PY - RX - qE + T$

subject to $G(Y^i, Q^i, X^i, E^i, Z, \theta^i) = 0$, θ^R is the level of θ reported by the agent through selection of a contract pair and $\hat{\pi}(Q, X, T; \theta)$ is the level of profits attainable if the agent does not participate in the programme. Conditions (a) and (b) ensure that a rent may be paid by the agency to encourage the agent's revelation of private information concerning the level of θ. While this represents an increase in profits, the function of this rent must be distinguished from one of income support. In fact, by design, eqn (8) ensures that the income increase is unrelated to such objectives and is limited to a level sufficient to induce conditions (a) and (b).

Alternatively, where agents hold some sort of altruistic preferences, the agency would be forced to take such preferences into consideration in design of the mechanism. In this case, the agency would solve the following problem:

$$\max S = E_\theta\{s[V(Q, Y) - PY] + (1 - s)U(Q, X, T; \theta)\} \quad (9)$$

subject to:

(a) incentive compatibility $U(Q, X, T; \theta, \theta) \geq U(Q, X, T; \theta, \theta^R)$

(b) individual rationality $U(Q, X, T; \theta) \geq \hat{U}(Q, X, T; \theta)$

where $U(Q, X, T; q) \equiv \max U(Q^i, Q\pi = PY - RX - qE + T)$

subject to $G(Y^i, Q^i, X^i, E^i, Z, \theta^i) = 0$, θ^R is the level of θ reported by the agent through selection of a contract pair and $\hat{U}(Q, T; \theta)$ is the level of profits attainable if the agent does not participate in the programme. The need to consider producer preferences becomes clear when the conditions of eqn (8) (a) and (b) are reconsidered when agents are non-hedonistic. That is, in order to ensure that the agents are left better off participating in the programme, the agency must design the programme subject to the constraint that agents' objectives are improved, not necessarily that just profits increase. The importance of the possible role of producer preferences has often been noted in discussions of the role of producer 'goodwill' in determining participation (see, for example, Whittaker *et al.*, 1991). Here, the policy approach is able to take these preferences into consideration and provide a mechanism that is compatible with private, voluntary provision of the environmental goods.

The solution to these problems provides menus of contracts that vary with the level of quasi-fixed factors managed by the agent. The agency offers the agent the opportunity to select a contract that maximizes private objectives. By design, the menu leads the agent to find it optimal to choose the contract associated with the true level of the agent's quasi-fixed factors. Most importantly, this approach does not require that the agency knows the level of the quasi-fixed factors, only their distribution across the population of producers. Thus, in comparison with the direct payments-for-standards approach, the individual management-agreement approach represents regulation that allows the agent to select the prescription that is optimal, given the agent's endowment. This provides the basis for a substantial gain in efficiency. Importantly, this approach should not be confused with individually negotiated management agreements where the agency is presumed to have no knowledge of the underlying technology. In the case considered here, individual negotiations are completely avoided by the use of a mechanism design that induces agents to reveal their true quasi-fixed factor endowments. In comparison, this advantage is gained at the cost of information rents, defined by $U - \tilde{U}$ and $\pi - \tilde{\pi}$.

CONCLUSIONS

This chapter has presented a new perspective on the question of public policy appropriate for encouragement of the voluntary, private provision of environmental public goods. While the results are of particular interest for agriculture, they are applicable to other sectors as well. The second section of the chapter clarifies the rationale for a public-sector role. This is particularly relevant for agriculture. It is argued that agriculture involves a technology that is joint in private and environmental public goods. This, together with the spatial scope and diversity of agriculture, suggests that direct public provision of environmental public goods is not feasible. The chapter then shows that, while investment grants may be a useful mechanism for public control, the approach is conceptually equivalent to the use of individual management agreements that operate over variable private goods. In comparison, the approach of uniform direct payments for uniform standards results in substantial inefficiency, which results from the use of a uniform policy across heterogeneous agents, as well as from the need to select a uniform prescription. The uniform prescription results necessarily in inefficiency associated with the need for agents to adjust their production plan to the prescription. This inefficiency is most notable for agents that may be producing levels of environmental public goods which are socially optimal, though with production plans that otherwise do not conform with the uniform standard. The policy inefficiency results from its requirement that such agents reallocate private goods to the evidently inefficient policy prescription.

The alternative approaches can also be compared with respect to the level of technical distortions and rents they produce. Going beyond the technical distortions of uniform standards already noted, it is well known that any standards approach or marginal incentive scheme would also distort other private-good choices from their private and socially optimal levels. The advantage of the individual management-agreement approach is that the menu of contracts is designed to ensure that resulting production plans are optimal, given the agency objective. While private choices are changed, they are shifted to a new socially optimal expansion path. On rents, the chapter shows that uniform direct payments for standards are likely to result in substantial rents paid to producers, compared with those paid through the use of individual management agreements. This reduction in rent results from a design that leads agents to self-select a management-agreement contract that is consistent with their true level of quasi-fixed factors. Thus, while rent is paid for this revelation, no further rent is paid as a result of the agency being incorrectly informed of the quasi-fixed-factor endowment.

NOTES

1 Archibald (1988) and Weaver and Harper (1993) noted the importance of integrating private-good and biophysical processes, allowing for input jointness, as in eqn (1). More generally, Meade (1952) presented an early discussion of jointness inputs observed in the production of externalities, while Baumol and Oates (1975) specified externalities as produced in fixed proportion with private inputs. Sudit and Whitcomb (1976) used a more general joint-input specification. In each case, these restricted forms of eqn (1) allow the producer to choose private production plans without consideration of external effects.

2 The implications for private choice of the functional properties of the technological interaction of the private and public goods are important. By definition, private choices based on the myopic production function will be both privately and socially inefficient when a joint production function such as eqn (1) applies.

3 We use the standard terminology here. We have labelled the individual decision-maker as the agent and distinguish this agent from the public organization (labelled agency) attempting to control individual decisions. This usage is intuitive, since the agency attempts to control the agent.

Mechanisms for the Provision of Public Goods in the Countryside

12

Uwe Latacz-Lohmann

INTRODUCTION

Recent decades have seen a continuous increase in public demand for rural environmental goods and services, such as aesthetic quality of the rural landscape, wildlife habitats or recreational opportunities. Over the same period, the supply of such benefits has been seen to be in decline. It has been argued with increasing vigour that many of the environmental and cultural functions of the rural environment have been subjected to growing threats. The proximate cause of this conflict is changes in agricultural practices, induced by a combination of agricultural price support and technological progress. Behind them lies a more fundamental cause, which economists refer to as market failure. The inability of the market to lead the economic process to an optimum, that is to say, to match demand with supply, relates back to the public-good character of most of the environmental goods and services concerned. The major part of the potential value of countryside benefits accrues to people not involved in their production (Hodge, 1991). As it is generally not feasible to exclude those who have not contributed to the costs of their provision from using and enjoying the benefits, private producers have difficulties charging for the use of public goods and thus may be reluctant or unwilling to invest in their production. For this reason, private markets tend to undersupply public goods relative to efficient levels. It is therefore widely accepted that some institution other than conventional markets is needed to stimulate the provision of public goods.

This chapter provides a theoretical analysis of three quasi-market mechanisms that have been used to stimulate the provision of countryside benefits by farmers: posted prices, auctions and individual negotiations. All

 The Economics of Landscape and Wildlife Conservation (eds S. Dabbert, A. Dubgaard, L. Slangen and M. Whitby)

three mechanisms offer positive financial incentives for the provision of environmental goods and services by private enterprises, although the mechanisms differ in the way these incentives are arranged. This policy of positive incentives is widely applied and well established across the European Union (EU), its key legislative tool being the management agreement.

THE PAYMENT MECHANISMS DEFINED

Posted Prices

Posted prices (fixed-rate payments) are the simplest and most widely used mechanism. An 'adequate' price for the good (in terms of incentive payment) is decided centrally by the government and then forwarded to the farmers. Most European environmental conservation and enhancement programmes, as technological generators of public goods, operate on the basis of this mechanism. Participating farmers receive predetermined flat-rate payments for following prescribed management practices. In the UK, it was the launch of the Broads Grazing Marshes Conservation Scheme and the development of the Environmentally Sensitive Area (ESA) schemes that increased the importance of this payment mechanism (Fraser, 1992). Although farmers can choose between packages (so-called 'tiers') of management restrictions of varying stringency and payment levels, the payment level within each tier is predetermined and equal for all participants. Typically, payment rates are derived from the profit forgone of a representative farm. They therefore reflect the average opportunity costs of establishing the conservation practices concerned.

Auctions

The award of contracts on the basis of competitive bidding is a method frequently used in procuring commodities for which there are no well-established markets (Holt, 1980). An auction market for public goods in the countryside could be set up in the following way. The environmental agency puts conservation contracts up for tender. The contracts must be clearly specified as to the management prescriptions they involve. Farmers, as potential providers of the services, submit bids to the agency indicating the amount of payment that each of them would require to adopt the conservation practice in question (Latacz-Lohmann and Van der Hamsvoort, 1996). The first-price sealed-bid auction is the most suitable auction form to be used in this context.[1] Potential providers tender written bids within a stipulated time, after which the bids are opened and compared. Typically, in a competitive sealed tender, the contract is awarded to the lowest bidder (Friedheim, 1992). 'Green auctions', however, differ from the standard auction in that multiple similar contracts (rather than one unique contract) are offered. Under this auction regime, the *n* lowest bidders are rewarded, receiving the payment stated in their bids (discriminatory first-price sealed-bid auction). In practice, the operating agency may set a threshold that successful bids must not exceed. The US Conservation Reserve Program (CRP) provides us with a working example of this policy. Since 1986, the US Department of Agriculture has awarded contracts for this programme on the basis of a competitive bidding mechanism.

Negotiations

Negotiations between individual farmers and conservation authorities have been used to determine appropriate prices for the provision of environmental benefits on Sites of Special Scientific Interest (SSSI) in the UK. This approach is very similar to the Coasean bargaining solution, the major difference being that management-agreement bargaining takes place in an environment of asymmetric information. Also, the environmental agency has important powers of compulsion, which may be brought to bear to encourage reluctant farmers to reach an agreement (Withrington and Jones, 1992).

CRITERIA FOR EVALUATION

This section specifies what might be meant by 'market performance' or 'success' in the context of public-good provision. Although several criteria for policy evaluation have been proposed in the theoretical literature (see, for example, Field, 1994), we shall focus on the one that appears most appropriate for the evaluation of quasi-markets: cost-effectiveness. A cost-effective public-good service is one that minimizes the costs of service delivery. Depending on what kind of cost is considered, two variants of cost-effectiveness can be distinguished.

Cost-effectiveness Based on Economic Costs

The overall economic costs of public-good provision comprise agricultural opportunity costs and transaction costs. Opportunity costs embrace all direct and indirect farm-level costs associated with the adoption of the conservation practices concerned. They are most aptly described as 'farming profits forgone', valued at social costs plus any direct costs. Transaction costs can be divided into two kinds: *ex ante* and *ex post* exchange (Williamson, 1975). *Ex ante* transaction costs are the costs incurred in preparing, negotiating and safeguarding a conservation agreement. These include, among others, the costs of agri-environmental data collection and the time and resources used in negotiating the agreement or, in the case of auctions, the costs of bid preparation and implementation. *Ex post* transaction costs are the costs of monitoring the farmers' compliance with the terms of their contracts and any costs of conflict resolution if the terms have not been complied with (Le Grand and Bartlett, 1993). Note that transaction costs mainly arise from the need to deal with informational deficiencies and asymmetries. Although both sides of the market are affected, the bulk of transactions costs are borne by the operating agency in the form of administration costs. The three incentive mechanisms analysed in this chapter vary mainly in *ex ante* transaction costs.

Cost-effectiveness Based on Budgetary Costs

In the political arena, the cost to the exchequer is often considered more important than economic cost. The budgetary cost elements in this system are those of the conservation payments and the operation of the quasi-market (that is to say, administration costs), corrected for indirect cost savings following adjustments to farm output (Whitby and Saunders, 1996). Usually, the environmental agency has a fixed amount of public money available for issuing conservation contracts, and the agency would try to maximize results within the overall budget. This is an exercise in public finance rather than one in welfare economics.

How is budgetary cost related to economic costs? Budgetary costs will be higher than economic costs because, with price support, profits forgone by farmers (on which incentive payments are based) exceed the real value to society of farming profits forgone (Whittaker *et al.*, 1991). At the margin, economic costs of the contracts may even be negative, if inefficient surplus production is being removed. Furthermore, farmers are likely to earn informational rents in excess of farming profits forgone.

Consideration of Quality

As conservation contracts may differ widely in the quality of the public goods they generate, some indicator of quality should be included in the cost-effectiveness measure. Ideally, a contract would specify a clearly defined improvement in environmental quality. However, because of difficulties in measuring environmental quality directly, the quality indicator would have to be an observable technological proxy, such as the difference between initial and subsequent land-use intensity. Given the availability of this information, the costs of each contract could be related directly to the quantity and quality of the public good generated.[2] Following Le Grand and Bartlett (1993), I shall refer to this extended concept of cost-effectiveness as productive cost-effectiveness, which, in less technical jargon, may best be described as 'value for money'.

A THEORETICAL ANALYSIS OF QUASI-MARKET MECHANISMS

In this section, theoretical models of the three incentive mechanisms are presented. Emphasis is placed on the farmers' responses to the incentives provided by each of the mechanisms and on the resulting 'price' of the service. The results will be discussed in the light of the above evaluation criteria.

Posted Prices

Assuming profit-maximizing behaviour, farmers will be willing to participate in a fixed-payment conservation scheme only if the payment offered at least covers their opportunity costs, that is to say, profits forgone. In the terminology of supply theory, opportunity costs could be seen as the variable costs of public-good provision, as they increase with the quantity and the quality of the service delivered. Similarly, transaction costs incurred by the farmer could be considered fixed costs of public-good provision, as they are normally incurred as a lump sum independent of the level of service delivery. They embrace the costs of gathering information about the conservation scheme, the costs of assessing the technological and financial consequences of participation (for example, consultancy costs), the costs of connecting and interacting with the environmental agency, and so on.

The upper chart of Fig. 12.1 shows a simple model of public-good provision under a fixed-rate offer system. The industry supply curve, denoted S, has been drawn on the assumption that participation in a conservation scheme, for example, within an ESA, is offered at increasing marginal costs. This reflects the fact that it becomes increasingly costly to put higher-quality land under a conservation agreement. Note that the *ex ante* transaction costs incurred by the farmers do not affect the supply curve, because they are sunk costs at the time the decision about participation is made. However, if these costs are perceived to be unreasonably high they may deter farmers from considering participation.

Note that 'supply' in the top chart of Fig. 12.1 refers to the land area that farmers are willing to commit to the conservation scheme, and not to the ultimate product 'public good'. Under a fixed-rate payment scheme, with the fixed payment denoted $\bar{p}$, farmers will be willing to enrol land between points 0 and B. To the right of point B, participation is not worthwhile because the costs incurred would exceed the payment offered.

The bottom chart of Figure 12.1 provides a conceptual framework to take account of differences in the quality of public good delivery on different tracts of land. The two curves labelled EQI (environmental quality index) indicate the level of environmental quality that would be attained with conservation agreement (index 1) and without (index 0). The EQI could be thought of, for example, as the number of rare species per hectare. The vertical distance between the two EQIs indicates the quality of the public good (for example, regional biodiversity) that each individual hectare would provide if enrolled in the scheme.

Figure 12.1 has been drawn on the assumption that the quality of the public good provided is a positive function of the costs incurred in its production. Provision of high-quality benefits requires significant changes in land use and management practices, which imply high opportunity costs. Moderate

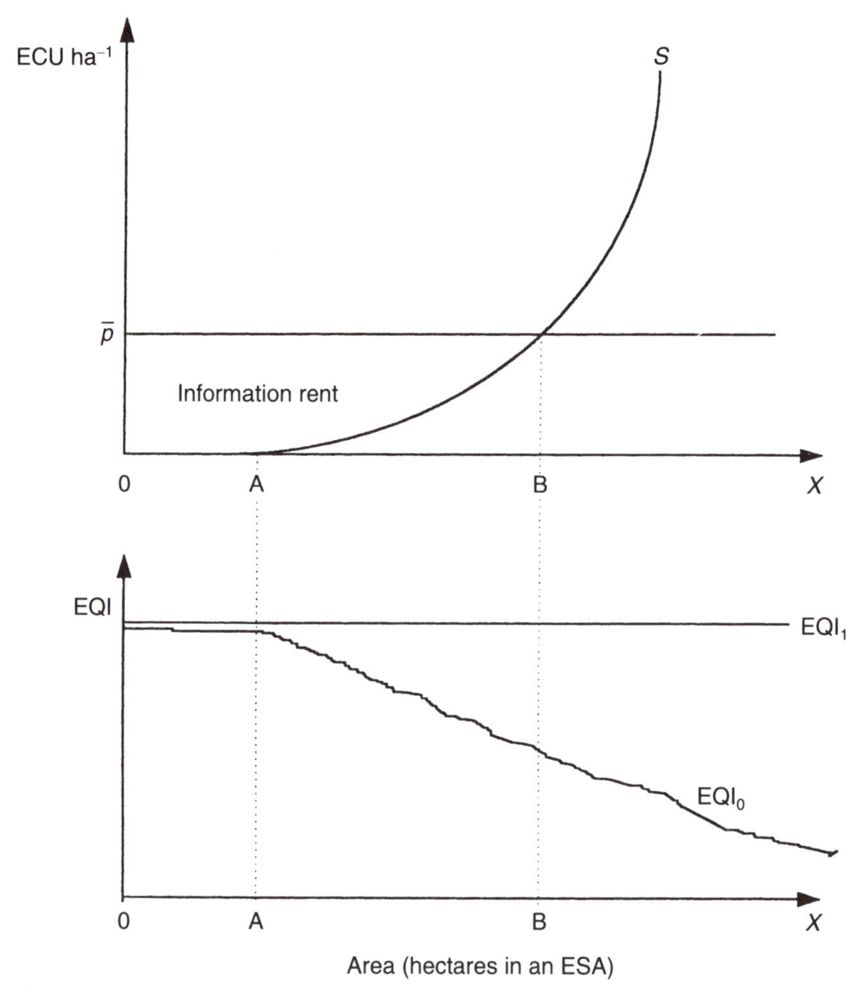

Fig. 12.1. Public-goods supply at the industry level under a fixed-rate payment scheme. ECU, European currency unit.

changes in land use will yield only moderate benefits and, when no changes are necessary at all, there will be no improvement in environmental quality. This is the case for land between points 0 and A, where the conservation agreements only safeguard the current status of the environment without adding value to it. Note that, under the assumptions made, it is not possible with a standard-rate payment scheme to target land to the right of point B, which would provide comparatively high environmental benefits if enrolled.

The area between the price line and the supply curve in Fig. 12.1 depicts what in conventional markets would be termed producer surplus. In the context of a quasi-market, it is more aptly described as informational rent, because the agency lacks the information to specify individual rates equal to opportunity costs. As is clear from the figure, free-riders (between points 0 and A) and low-cost, low-quality producers secure the highest economic rents, while potential high-quality producers (to the right of B) are discouraged from participation. In the information-economics literature, this is described as adverse selection. Adverse selection occurs where the condition of the transaction (that is to say, information asymmetry) attracts to one side of a transaction individuals with characteristics that adversely affect the value of the transaction to the other party (Williamson, 1985). In this case, information asymmetry leads to a selection of farmers and land with a low potential for environmental improvements. This is a significant drawback of the fixed-rate payment system.

Auctions

Auctions have several theoretical advantages over standard-rate payments. Firstly, they explicitly introduce an element of competition between the producers. Producers facing a competitive environment are likely to compete away, at least partly, their informational rents. This enhances (budgetary) cost-effectiveness. Secondly, auctions enable the participants to deal with information asymmetry and the uncertainty about the value of the (non-market) object being traded (McAfee and McMillan, 1987). Thirdly, auctions allow for consideration of differences in the quality of the service, in that the bid-selection process can give priority to those bids that offer the highest potential for environmental improvement per monetary unit of bid. Deeper insights into the performance of public-good auctions and, in particular, into price formation in the auction market can be gained by modelling bidding behaviour.[3]

The starting-point of the analysis is the assumption that the bidders' strategies are guided by the notion of a maximum acceptable payment level b, a reserve price above which no bids will be accepted. Of course, the bidders do not know this bid cap *ex ante* and will therefore form expectations about it. These can be characterized by their density function $f(b)$ and distribution function $F(b)$ over the range of possible outcomes. This range is bordered by $\underline{\beta}$ and $\overline{\beta}$, the minimum and maximum expected bid cap, respectively. The probability that a given bid is accepted can then be written as:

$$P(b \leq \beta) = \int_{b}^{\overline{\beta}} f(b)\mathrm{d}b = 1 - F(b) \tag{1}$$

where b denotes the actual bid price submitted. The bidder's decision problem now is to determine the bid that maximizes the expected net payoff from participation in the conservation scheme. Expected net payoff is the additional income from participation times the acceptance probability:

$$(\Pi_1 + b - \Pi_0) \cdot (1 - F(b)) > 0 \tag{2}$$

where Π_0 and Π_1 denote profits from farming (without conservation payments) under the conventional and the conservation technology, respectively. The optimal bid is found by maximizing expression (2) with respect to b. This yields:

$$\mathrm{b}^* = \Pi_0 - \Pi_1 + \frac{1 - F(b)}{f(b)} \tag{3}$$

Note that *ex ante* transaction costs (in this case, the costs of information and bid preparation) do not enter the optimal-bid considerations, because they are sunk costs once the bid is being submitted. However, these costs imply a loss to the farmer if the bid is rejected and a reduction in the accruing rent if the bid is accepted.

If it is assumed that the bidder's expectations are uniformly distributed in the range $[\underline{\beta}, \overline{\beta}]$, the optimal-bid formula becomes:[4]

$$b^* = \max\left\{\frac{\Pi_0 - \Pi_1 + \overline{\beta}}{2}, \underline{\beta}\right\} \text{ subject to } b^* > \Pi_0 - \Pi_1 \text{ (participation constraint)} \tag{4}$$

The participation constraint ensures that only cost-covering bids are submitted.

Expression (4) reveals important information about optimal bidding behaviour and thus price formation in an auction market:

- The optimal bid is a linearly increasing function of the bidder's opportunity cost of adopting the conservation technology. This implies that the bid conveys information about the bidder's cost type. A high bid signals high opportunity costs, and vice versa. This cost-revelation mechanism reduces the degree of information asymmetry and thus diminishes the informational rents accruing to providers.
- However, there is still a free-rider problem, in that farmers who have already been applying the conservation technology, and therefore incur no additional opportunity costs when implementing the conservation contracts, will submit a positive bid of $\frac{1}{2}\overline{\beta}$ (or at least $\underline{\beta}$).

- Price formation is also determined by the bidders' expectations about the maximum acceptable payment level. The optimal bid increases linearly in $\overline{\beta}$. Note also that a rational bidder will never bid below the lower threshold $\underline{\beta}$, because any undercutting of this level would not increase acceptance probability, while it does reduce net payoff. This implies that optimistic expectations will lead to higher budgetary costs of public-good provision.

Figure 12.2 provides a graphical representation of price formation in an auction market for countryside benefits. The optimal-bid curve has been calculated, based on eqn (4), on the assumption that the bidders' expectations about the maximum acceptable payment level are distributed in the range of minus 30% ($\underline{\beta}$) to plus 30% ($\overline{\beta}$) of the average opportunity cost of switching farming practices.[5] If, for example, the average cost is 100 European currency units (ECU) ha^{-1}, the farmers expect the cut-off level for bid acceptance to lie somewhere between 70 ($\underline{\beta}$) and 130 ($\overline{\beta}$).

As can be seen from Fig. 12.2, the informational rents accruing to producers are smaller than under the fixed-rate payment mechanism, implying a higher cost-effectiveness in terms of public money. Overall expenditure under the auction mechanism is reflected by the area underneath the bid curve between points 0 and C. The corresponding figure for the standard-rate payment scheme is the area under the price line $\overline{p}$ between points 0 and B. The graph shows that, with approximately the same amount of public money, more land can be attracted into the scheme when an auction mechanism is employed.

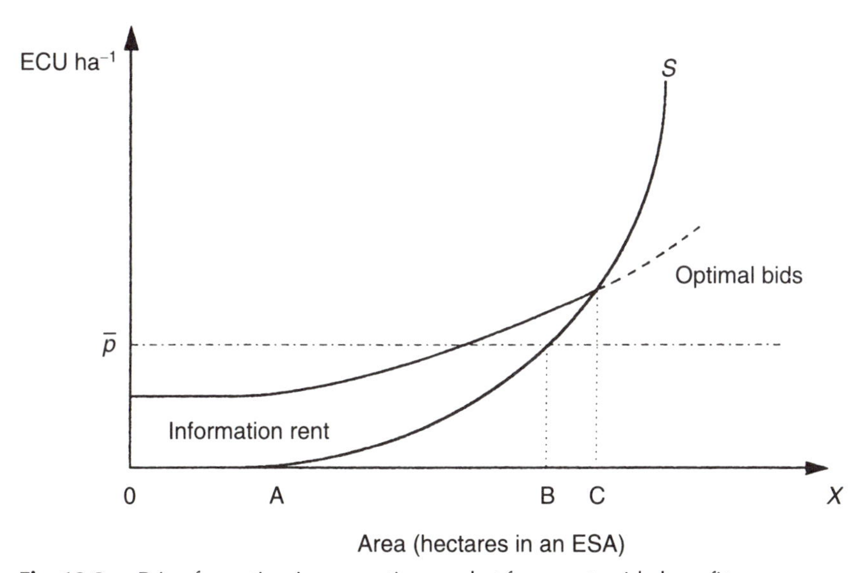

Fig. 12.2. Price formation in an auction market for countryside benefits.

Recall from Fig. 12.1 that this additional land (between points B and C) provides, on average, a higher quality of service delivery than the land to the left of point B. This implies that the productive cost-effectiveness (average budgetary cost per unit of environmental improvement) under an auction mechanism is even higher than the cost-effectiveness based on the average cost per hectare of land enrolled.

The performance of the auction could be enhanced further if the conservation potential of each individual bid were explicitly taken into account in the bid-selection process. Ideally, the operating agency would rank all bids for acceptance based on the ratio of expected environmental benefits to the public cost of the contract. However, as indicated earlier, this requires information about agri-environmental relationships at the farm level, which may not be consistently available. Model calculations reported in Latacz-Lohmann and Van der Hamsvoort (1996, 1997) indicate that the efficiency gains that may be achieved by directly targeting high-quality services are significant and may in fact call for increased investment in agri-environmental data collection.

Finally, two things should be noted. Firstly, the operation of an auction, whether with or without cost-effectiveness targeting, is administratively more difficult than the operation of a fixed-rate payment scheme, implying higher transactions costs. Secondly, US experience with the CRP auction has shown that multiple bidding rounds may give rise to decreasing bidding competition. After several sign-ups, farmers had learned the bid caps, which resulted in the majority of bids being almost exactly equal to the bid cap (T. Osborn, 1994, personal communication). This has made the potential benefits of the auction largely vanish and has given rise to massive criticism (US General Accounting Office, 1989).

Individual Negotiations

Individual negotiations between farmers and an environmental agency over an appropriate price for the provision of environmental services may be regarded as an application of the Coasean bargaining model. The strong form of the Coase theorem states that voluntary negotiation will lead to a Pareto-efficient outcome. Fraser (1995) has shown, however, that management-agreement bargaining is unlikely to achieve efficiency, because of the presence of information asymmetry. Without full information, the environmental agency is unable to assess the farmer's opportunity costs and thus does not know the farmer's pay-off as a function of the strategies pursued. The bargaining process therefore begins with a high degree of uncertainty on the side of the agency.

Fraser (1995) develops his arguments along the lines of a bargaining process that begins with an offer from the environmental agency, which the farmer may either accept or reject (that is to say, there are no counteroffers). If the farmer rejects the offer, this may be for two possible reasons. Firstly, the offer does not cover opportunity costs. Secondly, the offer may indeed be cost-covering, but the farmer is hoping for a higher offer in the next negotiation round. In the latter case, the farmer is effectively trying to exploit his/her informational advantage to maximize the economic rent. Because of informational inadequacies, the agency is unable to establish the true reason for the rejection of the offer. Given a high enough valuation of the contract, the agency may choose to increase the offer in the second round, and the farmer again may either accept or reject. The game might be continued in this way over several rounds, and a rational producer's strategy would be to reject any given offer if he or she expects a higher offer to be forthcoming in the next round. Only in the last round would the farmer have a dominant strategy of accepting any last 'take it or leave it' offer that exceeds his/her opportunity costs (Fraser, 1995).

Three conclusions may be drawn from this. Firstly, there is a positive probability that no agreement is reached, which means that the gains from the potential trade are lost. This may particularly be the case if the agency underestimates the other party's true opportunity costs and thus fails to offer a high enough price in the final round. A positive probability of disagreement should, however, not be considered a particular drawback of negotiations, because this is also a possible outcome under the other two mechanisms. Secondly, if true opportunity costs are overestimated or the environmental agency employs a soft bargaining strategy in order to avoid disagreement, producers receive an informational rent. Thirdly and most importantly, the negotiation process provides the farmers with strong incentives for blackmail, in that low-cost producers may disguise themselves as high-cost producers in order to qualify for higher payments. It is this excessive form of adverse selection that distinguishes negotiations from the other two mechanisms. The possibility of a low-cost, low-quality producer securing a contract designed to suit a high-cost, high-quality provider opens scope for a high degree of inefficiency – an outcome that would not be possible under an auction mechanism or a fixed-rate payment scheme.[6] As Fraser (1995) rightly points out, 'If farmers wished to be unscrupulous in the calculation of compensation payments, they could be' (p. 24).

This raises an obvious problem for the environmental agency. On the one hand, the agency needs to be prepared to reject conservation proposals that involve unnecessarily high levels of compensation. A tough bargaining strategy would gain the agency a stronger reputation and would reduce the likelihood of blackmail. On the other hand, a strategy that uses the possibility of disagreement as an explicit tool of the negotiation process is at odds with the agency's task of providing public goods and services in sufficient quantities. The agency must therefore strike a balance between the need to perform its task sufficiently and the requirement to remain within the designated budget (Fraser, 1995).

There are essentially two forces that may induce producers not to protract the negotiation process and to accept a cost-covering, but not excessive, offer before the final round. The first is impatience. Any rejection of an offer means a delay in the receipt of the net pay-off from the trade (Kennan and Wilson, 1993). This implies that the presumably higher surplus that is expected to be forthcoming in round $t + 1$ must be discounted to make it comparable to a given offer in round t. Hence, the producer might accept a less attractive offer in t when the latter, *ceteris paribus*, exceeds the present value of the expected payment in $t + 1$. This mechanism is, of course, only effective if a long enough period of time lies between the negotiation rounds. On the other hand, impatience on the part of the conservation agency will induce the agency to increase its offer in period t in order to avoid delay.

A second force that might prevent excessive delay is the cost of continuing the negotiation. Each party may see its costs of protracting the negotiation process as not worth the possible prize (Kennan and Wilson, 1993). Hence, negotiation costs may act as an incentive to compromise. In this respect, negotiation costs actively influence price formation.[7] Whitby and Saunders (1996) note that, in the SSSI negotiation process, farmers entering into agreements may claim reimbursement of negotiation costs to avoid such costs becoming a deterrent. Despite the good intention, this arrangement is likely to aggravate the problem of protraction and contribute to the inefficiency of the negotiation process.

Notwithstanding these drawbacks, one should not lose sight of the possible advantages of the negotiation mechanism. Bargaining is substantially a process of communication between the two parties, necessitated by differences in information known to the parties separately. The process of bargaining, though costly, can be an efficient way for the parties to get to know one another better, to build good working relationships and to establish a common informational basis for an agreement (Kennan and Wilson, 1993). Each party tries to infer from the other's move the other's true valuation and updates his/her beliefs when new information becomes available. Thus, delay may be seen as a tool for conveying private information credibly. Through this process of Bayesian learning, the informational imbalance may be rectified and farmers may be induced to act more honestly.

Another advantage of negotiations is the possibility of responding flexibly to specific needs and preferences of consumers. For example, the environmental agency can adjust the management prescriptions to local environmental conditions and conservation requirements. Such spatial targeting is generally not feasible under an auction mechanism, because the operation of this mechanism requires the auctioning of a large number of identical contracts in order to make the competing claims comparable. Another important aspect in the context of quality targeting is the possibility of accepting high-cost, high-quality producers. This is only to a limited extent possible under an auction scheme and impossible when a fixed-rate payment mechanism is used.

SOME BRIEF CONCLUSIONS

The analysis in this chapter suggests that the relative preferability of the three incentive mechanisms depends crucially upon the availability of information to the operating agency. Auctions are the most powerful mechanism for the agency to deal with severe informational deficiencies and asymmetries, because the cost-revelation mechanism inherent in the bidding process partly rectifies these imbalances. Individual negotiations may be expected to yield efficient outcomes when a common informational basis for an agreement already exists or when the necessary information is easy to obtain and assess. Standard-rate payments take a middle ground between these two. Their performance is independent of the availability of information, as the existence of informational rents is an accepted element of this mechanism, and no attempt is made to deal with this problem (Fraser, 1992). In an environment of severe informational asymmetries, fixed payments outperform negotiations, because they reduce the scope for opportunistic behaviour, but they are inferior to auctions, because they fail to exploit the efficiency potential of competition between producers. In any case, they offer savings in transaction costs. When a good informational basis exists, standard payments are outperformed by the other two mechanisms, because they fail to make use of the information available. Bearing these arguments in mind, auctions should be employed when the informational basis is weak, the number of potential participants is large and the contracts offered are homogeneous but the farms are heterogeneous. Standard-rate payments are suitable under the same circumstances, except that the farms should be homogeneous. The appropriateness of negotiations appears to be restricted to individual cases of site protection that require specifically tailored management prescriptions and involve only a limited number of farmers and good information.

NOTES

1 There are many other auction forms such as the English, Dutch, second-price sealed-bid and double auction. For a classification of these and their applicability to conservation contracting, see Van der Hamsvoort and Latacz-Lohmann (1996).

2 'Quantity' in this context could be thought of as the number of hectares put under conservation agreement. 'Quality' refers to the actual environmental improvements achieved on these hectares.

3 The following is a summary of a bidding model developed by Latacz-Lohmann and Van der Hamsvoort (1997).

4 For a full derivation of the optimal-bid formula see Latacz-Lohmann and Van der Hamsvoort (1997).

5 This number has been chosen arbitrarily and only serves illustrative purposes.

6 Under the fixed-rate payment scheme, inefficiency is limited to the extent of low-cost, low-quality providers securing contracts that offer an average payment.

7 Recall that the corresponding transaction costs under the two other mechanisms (bid-preparation costs and information costs) are sunk costs and thus do not affect price formation.

Sociological and Economic Factors Influencing Farmers' Participation in Agri-environmental Schemes

13

Gottfried Kazenwadel, Bareld van der Ploeg, Patrick Baudoux and Gottfried Häring

INTRODUCTION

Recent agricultural practices have had an important influence on landscape and natural resources. This influence often reaches a level that threatens the sustainability of agricultural production at the farm level. As a result of Council Regulation (European Economic Community (EEC)) 2078/92, substantially different agri-environmental programmes have been developed and introduced in different member states and specific regions of the European Union (EU) to give incentives to farmers for a voluntary reduction of those farming practices which have a negative influence upon wildlife and landscape.

This analysis aims to examine the sociological and economic factors influencing farmers' willingness to participate in agri-environmental schemes and local initiatives for environmentally integrated farming (EIF). It is based on a substantial farm survey carried out in different regions of the EU. Methods such as inspection of tables and factor analysis were applied for analysing data. Many of the observations have the status of hypotheses for further research concerning these study areas or concerning cross-national analyses.

First, the main ecological targets and the main content of the EU schemes and other initiatives are described. Furthermore, causes for participation, partial participation and non-participation are explained. In addition, the endogenous development of EIF is also considered. Another key question addressed in the chapter is the extent to which local factors stimulate or hinder the growth of environmentally orientated farming from the farmers' perspective.

The results enable us to provide some recommendations for the improvement of the efficiency of the different schemes as to farmers' participation and for adapting them to the specific situation in the study regions.

THEORETICAL POSITION

For analytical purposes, it can be useful to distinguish an endogenous paradigm from exogenous theories of economic development. In endogenous theories, economic growth is determined mainly by qualities of economic systems (sectors, enterprises), whereas, in neoclassical theory, external conditions (technology, market) as so-called invisible hands show which course of development is most suitable (Romer, 1994). This reflects the distinction in the literature between the rational-choice theory (neoclassical) and agency theory (endogenous) of human behaviour (Zey, 1992).

Neoclassical mainstream research perceives entrepreneurs as rational decision-makers. Sophisticated researchers in this field, however, include endogenous factors, especially regarding the goals an entrepreneur holds and the range of alternatives he/she is open to. Special topics are levels of aspiration (maximizers versus satisfiers) and the handling of risk and uncertainty. Endogenous factors added often represent subjective factors (entrepreneurial goals and styles of farming reality), as opposed to objective external factors, such as prices of inputs and outputs. An illustration of an integrated approach is the subjective expected utility model of Lindenberg (1990).

Agency mainstream research especially about farming styles (van der Ploeg, 1994) perceives an entrepreneur as somebody who creates and defends his/her personal action space (room for manoeuvre). He/she does not in the first place make decisions on alternatives which appear from outside, but instead he/she is an actor working at a 'project' (enterprise). As an actor, he/she develops several kinds of resources, not only financial capital but also local knowledge (craftsmanship). Knowledge is interwoven with practical experience and it is developed step by step. Development is 'born from within' (Long and van der Ploeg, 1994). Sophisticated agency researchers, however, keep away from the illusion of 'everything goes'. This reflects the theory of structuration of Giddens (1984). Structure, such as farm structure and the local knowledge associated with it, always has two faces. It shapes possibilities (room for manoeuvre), but on the other hand it represents the limitations of the life world of farmers.

The section of this chapter about economic factors concerning farmers' participation in agri-environmental schemes has an exogenous analytical entrance. This is in line with the classical adoption–diffusion approach (Ruttan, 1996). Properties of individual entrepreneurs function as a filter, deciding on adoption or non-adoption. Figure 13.1 is a conceptualization of the approach. This contribution focuses on economic factors at farm level, extracting factors that hinder farmers' participation in agri-environmental programmes (intervening variables) and factors that push farmers towards participation (mediating variables).

This chapter later deals with farmers' endogenous development of EIF. EIF is defined as agricultural practices at farm level in which environmental goals (clean environment, wildlife conservation, landscape qualities, farm recreation and regional quality products) are integrated in agricultural production. This section has an endogenous analytical entrance and focuses on the farmers' perspective. From this perspective, farm factors and other local factors that stimulate or hinder the growth of EIF are examined.

The conceptual model (Fig. 13.2) can be read as an inventory of 'contingent conditions' to practices of EIF. The phrase 'contingent conditions' is borrowed from the realistic paradigm, as described by Drummond and Symes (1996) in a contribution about sustainable fisheries. They say that environmental policies should take into account the complex context of sustainable or unsustainable practices. Sustainable practices cannot be invented and managed top-down from the outside. Policy-makers should be aware of the

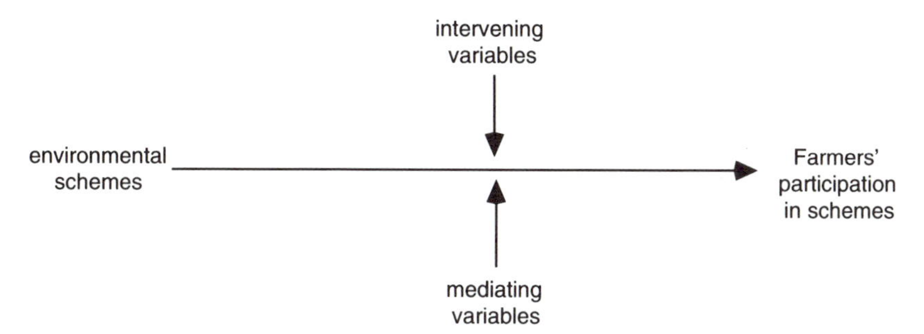

Fig. 13.1. Conceptual model of farmers' participation in agri-environmental schemes.

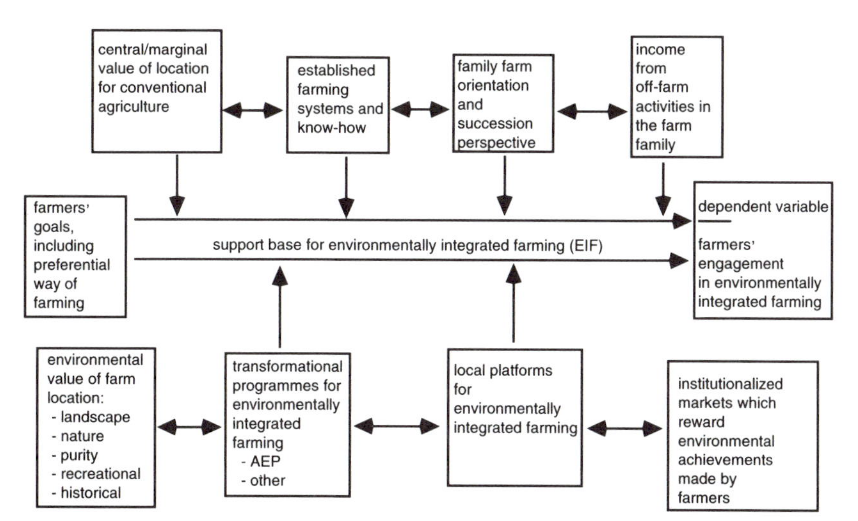

Fig. 13.2. Extended conceptual model of the farmers' live world in relation to the decision (not) to move towards environmentally integrated farming. AEP, agri-environmental programme.

shortcomings of 'environmental managerialism'. The same message is brought by Gray (1996) in his study on cultivating farm life at the borders, about sheep-farming in marginal areas. A keyword now is 'niche'. This is the equivalent of 'a whole of contingent conditions'. In hill sheep-farming, as well as in sea-fishing, the emphasis is on the creative improvisations of entrepreneurs and communities faced with specific natural, economic and political environments. Farmers and fishermen are active users of schemes and regulations. The way in which they do so can only be understood if their particular niche is taken into account.

MATERIAL AND METHODS

Material

This study is based on a farm survey carried out in different regions of the EU during 1995/96. In this chapter, the focus will be put on the results for the study regions in Germany and The Netherlands. In Germany the study regions are the Kraichgau and the Swabian Jura in Baden-Württemberg, and in The Netherlands Beemster and Waterland in Northern Holland.

In these regions, schemes under Council Regulation (EEC) 2078/92 have been implemented, but due to the different ecological conditions and political arrangements in the regions they have different targets and are implemented in different ways. The principal focus of the survey was farm structure and development, farmer's background, land use and labour input, the farm business, conservation investment and behaviour, future plans and expectations.

The sample in Baden-Württemberg was 150 farms in the two study regions and in The Netherlands 168 farms (including 48 in Waterland participating in schemes).

Methods

To evaluate the economic factors influencing farmers' participation in the schemes, the results of the survey have been analysed statistically. For Baden-Württemberg the farms are divided into participants and non-participants, and for The Netherlands for technical reasons in the following groups: participation in weak, middle and strong management agreements in Waterland. Therefore, only 48 farms of the whole sample of the 168 farms of The Netherlands sample can be considered. Statistical tests are used to evaluate whether these groups are significantly different concerning the tested parameters. For the survey data, a normal distribution cannot be assumed. Therefore, for quantitative data in the Baden-Württemberg survey the non-parametric Mann–Whitney U test is used. In Waterland, the Kruskall–Wallis test is used, because there are three groups to evaluate. Qualitative parameters are evaluated by the Pearson chi-square test.

The analysis later in the chapter of the endogenous development of EIF is guided by a conceptual model (Fig. 13.2). The theoretical background of this model is sketched in the second section. In qualitative field explorations in this study, this model can serve as a set of sensitizing concepts. In the quantitative analysis of survey data, the model can be interpreted in terms of variables of several orders (dependent, independent, intervening and mediating). First, farmers' goals are confronted with farmers' (non-) implementation of environmentally integrated agriculture. Factor analysis is used to extract different farming types in relation to EIF. Then, the opportunities in subareas for conventional farming are compared with the changes for wider (environmentally integrated) agriculture and for the specific farming systems in subareas.

STUDY REGIONS

The Agricultural Structure

The first Baden-Württemberg study region is the Swabian Jura, which is at an altitude of 400–1000 m above sea-level. It has varied environmental conditions. In the most favourable areas, high yields in cereals and silage maize can be obtained. But there are also large areas where low-intensity grassland dominates. Preventing contamination of water is important here, because this region contributes greatly to the drinking-water supply of central Baden-Württemberg. The Kraichgau is a region where natural conditions for arable farming are favourable (Table 13.1). Cereals, sugar beet, maize and oil-seeds

Table 13.2. Characterization of the study regions.

	Baden-Württemberg		Northern Holland	
	Kraichgau	Swabian Jura	Beemster	Waterland
Conditions for:				
Arable farming	Very good	Bad to good	Very good	Bad
Grassland farming	Bad to good	Bad to good	Very good	Average
Main farm types	Arable	Dairy Arable	Dairy Cattle/sheep Arable	Dairy Cattle/sheep
Main environmental values/problems	Erosion Species adapted to low-intensity land-use forms	Water protection Species adapted to low-intensity land-use forms	Cultural landscape Surface water	Bird protection Degradation of peat soils

are grown. Most of the farms are arable. Often the intensity of production is high. Due to the hilly landscape of the area, soil erosion is a major problem in the Kraichgau. In both areas, traditional low-intensity land-use practices tend to be given up and therefore the protection of species adapted to these practices plays an important role.

The Dutch subarea Beemster is very homogeneous and offers very good conditions for arable and grassland farming. Targets in this region are the preservation of the cultural landscape and the protection of the surface water. Waterland is more heterogeneous. Its peat soils are mostly used for extensive grassland farming (dairy, cattle and sheep). Bird protection and the degradation of the peat soils play an important role in this area.

The Agri-environmental Schemes under Council Regulation (European Economic Community) 2078/92 Applied in the Study Regions

Council Regulation (EEC) 2078/92 aims to remunerate farmers for their voluntary participation in programmes that help to conserve landscape and have an ecological impact. In this chapter, an overview of the agri-environmental schemes operating under this regulation in the study regions is given.

Schemes in Baden-Württemberg

In Baden-Württemberg, two schemes have been applied under Council Regulation (EEC) 2078/92: the market release and landscape conservation programme (MEKA) and the landscape preservation programme (*Landschaftspflegerichtlinie*, Ministerium für Umwelt Baden-Württemberg, 1990). The second scheme will not be considered in this chapter, due to its relatively minor importance.

The MEKA scheme is open only to farmers and has four main objectives (Ministerium für Ländlichen Raum, Ernährung, Landwirtschaft und Forsten Baden-Württemberg, 1993):

- Financial compensation for conservation and conversion to grassland in Environmentally Sensitive Areas (for example, areas in danger of soil erosion or nitrate leaching, and typical grassland landscapes).
- Protecting endangered land-use practices, such as using steeply sloping areas for grassland, traditional orchards and vineyards or low-intensity grassland.
- Remunerating extensive and organic arable cultivation.
- Conserving endangered biotopes.

The MEKA programme has been implemented across Baden-Württemberg with precise scheme characteristics, varying according to regional ecological problems. In 1994, it had a budget of 140 million Deutschmarks (DM), which was almost completely exhausted (Ministerium für Ländlichen Raum, Ernährung, Landwirtschaft und Forsten Baden-Württemberg, 1995, personal communication).

Schemes in The Netherlands

In 1992, Dutch management agreements (MMA) from the Ministry of Rural Areas, Nutrition, Agriculture and Forests came under Regulation 2078/92; they included several packages of restrictions and remuneration from which farmers could choose (Terwan and van der Bijl, 1996). Overall, there were three possibilities:

- Passive (weak) management agreements (so-called 'mountain area construction'), in which farmers are compensated for natural handicaps, especially wetland conditions;
- Light (middle) agreements, in which field management is made a secondary consideration in the interests of wildlife or landscape;
- Heavy (strong) agreements, in which severe restrictions are set for field management, in order to maintain botanical diversity in areas of peat pasture and to protect meadow birds.

Management agreements can be concluded only in certain zones, which cover about one-third of the area of one of the Dutch study areas, Waterland. There have been intensive discussions about the delimitation of these zones. In the subarea Waterland-West most of the land is under management agreements or rented out under restrictions by the state or conservation organizations, whereas in Waterland-East most of the land is outside the management-agreement zones.

Waterland is the focus for several other programmes for 'wider rural development' (farm recreation, regional quality products), with the focus agency at local and provincial level. The programmes are coordinated by a project bureau called Valuable Landscape (WCL) Waterland. Recently, attempts to encourage Beemster farmers to enter into wider rural development (in an 'administrative reallotment plan') have not been successful.

ECONOMIC FACTORS INFLUENCING FARMERS' PARTICIPATION IN AGRI-ENVIRONMENTAL SCHEMES

In this section the economic factors influencing farmers' participation in agri-environmental programmes are described and discussed. An overview of the results of this analysis is given in Table 13.2.

Table 13.2. Significance of differences between participants (P) and non-participants (NP), mean values and percentages of the researched parameters concerning participation in agri-environmental schemes.

	Baden-Württemberg (Swabian Jura and Kraichgau) n = 138 P n = 12 NP			The Netherlands (Waterland) n = 8 light MMA n = 18 middle MMA n = 22 strong MMA			
	NP	P	Significance	Light	Middle	Strong	Significance
Age (years)	52	47	NS	59	56	49	**
Family labour			**				NS
Total area (ha)	12	32	**	25	29	39	NS
Own area (%)	68	70	NS	61	46	44	NS
Farm type			NS				NS
Debts yes (%)	42	59	NS	38	44	73	*
Change of debts			NS				**
Income (%)			NS				NS
< 16,000	25	26		29	56	45	
16,000–32,000	33	46		57	44	28	
32,000–48,000	42	20		14	0	28	
48,000–64,000	0	5		0	0	0	
> 64,000 ECU	0	3		0	0	0	
Change of income			NS				NS
Investment yes (%)	33	74	**	50	61	86	**
Non-agricultural income (%)	79	51	**	61	40	26	**
Current situation (%)			**				NS
Excellent	0	1		0	0	5	
Good	8	18		33	39	33	
Fair	25	45		50	44	48	
Poor	58	29		17	6	14	
Bad	8	7		0	11	0	
Future situation (%)			NS				NS
Excellent	0	1		0	0	0	
Good	18	11		0	22	29	
Fair	9	35		83	50	48	
Poor	64	40		17	22	14	
Bad	9	13		0	6	10	

**, Significant at $\alpha = 5\%$; *, Significant at $\alpha = 10\%$; NS, not significant; MMA, management agreement.

Farm Capacities

Labour

For the family-labour parameter, no differences were found in Waterland. In Baden-Württemberg, family labour on farms taking part in MEKA is significantly higher than in other farms. While on MEKA farms family workers work about 3100 h $year^{-1}$, non-participants work only 1250 h $year^{-1}$ on their farm.

In Baden-Württemberg the use of hired part- or full-time workers is quite unusual. Only 4.7% of the participants in MEKA use hired non-family workers. Likewise, in Waterland hired labour does not play an important role. Only a few farmers (6.3%) employ hired part- or full-time workers, but the average is twice that of Baden-Württemberg. On average, there is no significant difference between participants and non-participants.

In Waterland, the age of the farmers corresponds significantly with the level of management agreements. Farmers participating in 'strong' schemes are younger than farmers only participating in schemes that do not need major changes on the farm ($\alpha = 5\%$). In Baden-Württemberg, there is no significant difference in the age of the participants and non-participants, but younger farmers tend to participate more intensively.

Land

In Baden-Württemberg, the total area of participating farms is considerably higher than that of non-participants. The average farm size of participants is 32 ha, whereas the average farm size of non-participants is 12 ha. This difference is significant ($\alpha = 5\%$). But there is no significant difference in the area owned by the farmers considered, although the average area of land owned by the participating farm family is more than twice as large as the area farmed by non-participants (18 ha and 8 ha, respectively). There is hardly any difference in the proportion of rented land between participants and non-participants. In Waterland, the total area of the farms between the three groups of management agreement increases with the degree of severity of management agreement, but the difference is not significant.

Capital

The capital capacity of the farms as such has not been a subject of the survey, but questions about the debt load and the change in borrowed capital have been included. Concerning the share of farms with borrowed capital, no difference between the groups can be detected, but in The Netherlands a smaller proportion of the farms participating in 'stronger' schemes tend to use more borrowed capital, although this is not significant ($\alpha = 10\%$).

In Baden-Württemberg, on many farms the share of borrowed capital has increased in the last 5 years, but there is no difference between participants and non-participants. In Waterland, particularly among farmers taking part in the most demanding management agreements, there has been a definite increase in borrowed capital during the last 5 years (significant for $\alpha = 5\%$).

An analysis of the investments made on the farms since 1985 (Table 13.3) shows that the share of MEKA participants who have made considerable investments (> 10,000 European currency units (ECU)) is higher than the share of non-participants ($\alpha = 5\%$). Possibly, this is also a result of the different farm sizes of participants and non-participants. The same relationship can be found in The Netherlands. There, the share of farmers who made investments over 10,000 ECU is higher in the group of farmers with more severe management agreements than the share in the group of farmers with less demanding agreements ($\alpha = 10\%$).

Farm Type

Farm types have been differentiated following the European farm type classification (dairy, cattle and sheep, pig and poultry, and so on; Communautés Européennes, 1985). Dairy, cattle and sheep farms have been divided into dairy farms and cattle- or sheep-producing farms. Neither in The Netherlands nor in Baden-Württemberg is there a significant difference between farm types and participation in schemes. Similarly, there is no significant difference between organic and conventional farmers regarding participation. This result is probably due to the limited number of organic-farming respondents.[1]

Economic Situation

The farmers were asked to specify their total family income before taxes. In Baden-Württemberg, there was a slight difference between participants and non-participants. Participants in MEKA tend to have a slightly higher income, but this difference was not significant. Similarly, in The Netherlands, the difference was not significant. In both study areas, there was no difference concerning income change in the last 5 years between the groups. The proportion of income from non-agricultural sources (off-farm work by respondent or other

Table 13.3. Percentage of farms with investments above 10,000 ECU in the last 5 years.

	Baden-Württemberg		The Netherlands		
Investments	Non-participants	MEKA participants	Light MMA	Middle MMA	Strong MMA
Yes	33	74	50	61	86
No	67	26	50	39	14

family members, pensions, tourism, investments and others) was higher for non-participants in MEKA and for participants in the weaker management agreements than for the other groups (Table 13.4).

The farmers have also been asked how they judge the current economic situation of their farm. In Baden-Württemberg, the non-participants judge their economic situation worse than the participants; in general, it is seen as poor to fair. In The Netherlands, there is no difference between the groups and the situation is judged fair to good. The future economic situation of the farms in the two study regions is always judged to be slightly worse than the current situation, but there are no significant differences between the study groups.

FARMERS' ENGAGEMENT IN ENDOGENOUS DEVELOPMENT OF ENVIRONMENTALLY INTEGRATED AGRICULTURE

Development of Environmentally Integrated Farming and Farmers' Goals

Indicators for our dependent variable 'development of EIF' are changes in farm structure or farm management which are chosen by farmers (among others) because of their positive environmental consequences and/or their strong market position due to specific local conditions.

Table 13.5 gives an impression of the relative number of farmers in the Dutch study area who were engaged in EIF during the last 5 years. The table also indicates how many farmers have such plans for the next 5 years. Changes which are not directly linked to the local environment are not included in the table. Also excluded are new off-farm jobs and the start-up of conventional second branches, such as horticulture on arable farms.

The most important targets of EIF are wildlife conservation (Waterland) and landscape conservation. In many cases, only modest changes in farm management have occurred, for instance, taking part in voluntary activities for the protection of meadowland birds (Waterland) or replanting old orchards (Beemster).

The responses to the question about future plans to participate in projects for regional quality products reveal a lot of uncertainty about this subject.

Table 13.4. Percentage of household income from non-agricultural sources.

Baden-Württemberg		The Netherlands		
Non-participants	MEKA participants	Light MMA	Middle MMA	Strong MMA
79	51	61	40	26

Table 13.5. Development of environmentally integrated farming in the past 5 years and future intentions for this activity.

Indicators for development of environmentally integrated farming	Percentage of farmers actually engaged in EIF or with plans to start EIF					
	Waterland					
	West (n = 54)		East (n = 59)		Beemster (n = 55)	
Past 5 years						
'Which of the following experiments of farm adjustments happened at your farm?'						
Wildlife conservation	15		12		2	
Landscape conservation	6		3		13	
Clean environment	2		3		7	
Organic farming	2		3		–	
Farm recreation	4		3		2	
Regional quality product	–		–		–	
*Next 5 years**						
'Which of these do you definitely plan to do, which might you consider?'						
To join a wildlife or landscape conservation organization	11	[30]	25	[54]	2	[9]
Extensification grazing	4	[9]	–	[8]	2	[6]
Less chemicals	5	[9]	8	[12]	2	[25]
To join a programme for organic farming	–	[2]	–	[10]	–	[2]
Farm recreation	2	[6]	2	[10]	2	[15]
Regional quality production	2	[20]	5	[24]	2	[18]

*Lower figure indicates 'plans', higher figure includes 'maybe'.

During the interview period, such products (for Waterland) were in a start-up phase. To a lesser extent, the same is true for farm recreation. Table 13.5 may underestimate the degree to which farmers are working in an environmentally friendly way. Changes to fulfil legal environmental obligations – particularly concerning manure and chemical inputs – are not included. Another question in the Dutch survey discloses that about 66% of the farmers in each of the subareas indicate that they are actively working towards a clean environment.

Farmers in the cross-national survey have been asked about the importance of specific goals (Table 13.6). Another question was about the perception of a hypothetical ideal farm: 'When you think of your ideal farm, which of the following characteristics call for your personal feeling of attraction or aversion?'

The results suggest that goals which are presented as independent variables in the conceptual model are not 100% independent. Selection processes can play a role. Some of the most ambitious entrepreneurs probably left the

Table 13.6. Goals of farmers and characteristics of an ideal farm.

	Percentage of farmers interviewed					
	Waterland					
	West		East		Beemster	
Goals*	(n = 54)		(n = 58)		(n = 55)	
Independence	72	[100]	71	[100]	84	[98]
Entrepreneurial challenge	7	[56]	34	[68]	35	[71]
Maximizing profit	13	[56]	14	[64]	24	[64]
Leisure time	19	[60]	12	[60]	15	[66]
Family farm tradition	13	[43]	18	[66]	13	[69]
Job satisfaction	91	[100]	86	[100]	82	[98]
Ideal farm†	(n = 54)		(n = 57)		(n = 54)	
Large	16	[30]	15	[28]	19	[47]
Intensive	22	[35]	23	[36]	22	[39]
Specialized	54	[59]	70	[77]	53	[72]
Mechanized	43	[59]	54	[67]	52	[70]
Organic farm	28	[41]	28	[54]	17	[41]
Free market products	14	[52]	37	[61]	35	[52]
Real family farm	67	[85]	70	[88]	61	[85]
Investments after savings	54	[63]	30	[53]	39	[52]
No activities other than agricultural production	94	[98]	82	[93]	91	[96]

*Lower figure (goals) = per cent farmers who answered 'yes, very important', higher figure includes answer 'yes, rather important'.
†Lower figure (ideal farm) = per cent farmers who feel attracted, higher figure includes 'neutral' (no aversion).

problematic subarea Waterland-West, and answering a question as to whether 'maximizing profit is important' may cause cognitive dissonance for farmers with few opportunities to realize any profits.

Farm characteristics most generally preferred are:

- Specialization in agricultural production.
- A 'real' family farm.
- Specialization in one main branch of agricultural production.

The general preference for specialized farming can mean that wider farm development is not the first choice of most farmers.

The profile of the Waterland-West farmers differs considerably from the profiles of farmers in both the other subareas. With regard to their ideal farm, they are relatively moderate in the preferences for specialized farming, mechanized farming and farming for free markets. Farmers in Waterland-East, however, have relatively high preferences for specialized farming.

Farmers in Waterland (West and East) tend more towards organic farming than their colleagues in Beemster. In Waterland-West this is connected with the feeling of almost being an organic farmer (extensive grazing), while Waterland-East farmers more often consider active organic-farm development as a real option. The large mental distance between Beemster farmers and organic farming cannot be isolated from their actual intensive way of farming.

The method of factor analysis with 'varimax' rotation was applied to variables mentioned in Table 13.6. This resulted in five factors:

1. Active entrepreneur: job satisfaction, independence, entrepreneurial challenge.
2. Active manager: mechanized, intensive.
3. Family farmer: real family farm, investments after savings.
4. Conventional farmer: no activities other than agricultural production, aversion to organic farming.
5. Calculating farmer: maximizing profit and leisure.

Farmers were classified in five groups on the basis of factor scores. It appears that farmers in two groups (active entrepreneurs and active managers) compared with farmers in other groups made much more investment in recent years. In Waterland-East, these farmers are often active members of a voluntary bird-protection group and the newly founded Nature Cooperative. The family-farm group is overrepresented in the subregion Waterland-West. Many of these farmers have most of their land in management agreements.

The classification in five groups was related to two EIF scores, a positive score (environmentally friendly farming) and a negative score (environmentally damaging farming). It appears that active entrepreneurs and active managers are relatively high at the positive and also at the negative ends of the scale. The three other groups (family, conventional, calculating), which do not change their farms considerably, were both lower at environmental benefits and damage.

Opportunities for Environmentally Integrated Farming

The realization of EIF requires a fusion between agricultural production values and environmental qualities. Waterland and Beemster are both areas with a long history of such a fusion. Waterland has an international status as a wetland area with unique natural values (species), bound to peat-meadow agricultural land use. Beemster has much less environmental interest but is clearly higher in landscape identity (unity) and in historical features. The area is on a list to be nominated as a United Nations Educational, Scientific and Cultural Organization (UNESCO) Cultural Landscape.

Neither Waterland nor Beemster has a tradition of exploiting markets for environmental qualities. Recreational activities have mainly been concentrated on the farmyards. Special agricultural products – with an environmentally friendly image – were quite exceptional, although both areas have large potential markets nearby for such products. Many parts of Waterland are located in urban-fringe zones. Because of its poor agricultural-production opportunities, Waterland has been urged to undertake a transformation toward economic exploitation of environmental qualities. Beemster farmers are proud of their beautiful polder – until now left intact by urbanization – but they do not consider it as a marketable product.

Established Farming Systems

Full EIF systems are marked by a double-sided relationship. Firstly, such systems contribute to the environmental qualities of the farm and its surroundings and, secondly, they obtain economic returns from environmental qualities. In Waterland-West, much land is owned by the state or conservation organizations – and rented out cheaply – or it is owned by farmers who have management agreements. Outside these management agreements, fully integrated farming systems are rare.

In Waterland and also in Beemster, most of the farms are too small to provide full employment and income for a farm family. Additional incomes are looked for in several directions. What is experienced as a potential is largely a 'social construction'. The social construction of environmental production – as an economic basis – became visible some 15 years ago in Waterland-West. This started mainly as a social construction from outside. Farmers who were in a difficult economic situation were offered management agreements. Later on, farmers in Waterland played a more active role in the social construction of the concept of environmental production, especially in relation to wildlife. It was used as a strategic concept to limit the area with a high water-table in Waterland-East, but the use of the concept did not only have strategic functions. Engagement in voluntary bird protection (Bouwhuis, 1995) probably stimulated internalization of the idea of environmental production as a part of the farmer's job. In this context, some 150 contracts concerning wildlife production have recently been concluded between the newly founded Nature Cooperative and individual farmers.

In Beemster, outside interference had quite different effects. Farmers in the polder areas have a strong feeling that they are the guardians of an attractive landscape. This awareness, however, has hardly any economic connotations. 'It should not be necessary to be paid for it' was an expression heard several times. Meanwhile, many farmers feel that additional income is needed. Commercial activities at the farm gate, such as recreation, are much closer to farmers in Beemster than wildlife production. Landscape maintenance is perceived as 'private' (farmyard) or as a by-product of rational agriculture.

CONCLUSIONS

General Impression

There are considerable differences in the degree to which farmers participate in Council Regulation (EEC) 2078/92. However, in both surveys there could not be found any group of farmers which is not interested in agri-environmental programmes or which rejects them totally. The involvement in environmental programmes and, more generally, in EIF can be rather passive but also very active. Impressions from the Dutch study areas indicate that EU schemes cover only a limited part of farmers' initiatives and goodwill regarding EIF.

Economic Factors Influencing Farmers' Participation in Agri-environmental Programmes

The Baden-Württemberg and Dutch results show that participation in Council Regulation (EEC) 2078/92 depends upon the farm size and whether the farm is run on a full-time or part-time basis. Small part-time farms often seem not to sign '2078' agreements (Baden-Württemberg) or to choose only weak management agreements (The Netherlands). In Baden-Württemberg, the main reason could be that expenditures are or are considered to be higher than the output from the scheme. In The Netherlands, further analysis may prove that marginal farmers are non-selective adopters: many sign weak management agreements for all their cultivated land.

Larger farms and farmers with a strong interest in an economic and efficient long-term agricultural production participate more often in agri-environmental schemes, especially in schemes which demand considerable investments to be made. In The Netherlands, larger farms often choose the stronger management agreements.

If a larger area is to be covered by the programmes, it will be necessary to improve the conditions for small farms in the study regions to find access to schemes or important parts of schemes. For economically orientated farmers, the subsidies of the schemes should be at least as high as the income losses; otherwise participation is not an option for them.

In disadvantaged regions farmers tend to be more in favour of EIF practices. The reason is probably the low range of necessary changes.

Endogenous Development of Environmentally Integrated Farming

There are also farmers who are producing in an environmentally friendly way but who do not participate at all in '2078' schemes. This does not always imply that EIF is recognized as a personal goal. An extensive way of farming may have unintended positive side-effects for environmental values, especially for biodiversity. Other farmers, however, are actively looking for possibilities to introduce aspects of EIF. An impression from Waterland is that especially farmers with relatively intensive farms who are most active in farm development are also most actively looking for EIF adjustment. Many farmers consider themselves as active supporters of the surrounding landscape and wildlife values. However, most of the time these initiatives are marginal in the mainstream development of their farms. The orientation of environmental values seems to be selective, at regional level and at the level of individual farmers. The marginal area in Waterland is more orientated to wildlife, while in Beemster farmers put a higher value on landscape.

Diversity and Scheme Design

Programmes that are designed to implement a certain level of EIF should take into account the different social and economic environment of farmers. The effects of actual programmes upon emerging initiatives of farmers can vary widely, from stimulating to blocking. It is important that EU programmes are differentiated by an adequate level of geographical scale. Further research will be needed to adapt schemes to the specific social and economic conditions in the target regions, so that, with a limited budget, EIF can be encouraged and the farmers can be remunerated for their environmental performances. Attention is also needed to the possible heterogeneity inside rural areas with regard to farmers' goals and farming styles.

NOTE

1 There are no organic farmers in the Baden-Württemberg survey who do not participate in the MEKA scheme, and in the Dutch survey there were too few organic farms to provide a statistically significant sample.

Optimal Allocation of Wildlife Conservation Areas within Agricultural Land

14

Ada Wossink, Clifford Jurgens and Jaap van Wenum

INTRODUCTION

Land use in most Western European countries is often characterized by apparently strong competition between agricultural needs and ecological requirements. Often competition is felt, while it must be recognized that understanding the ecology of the landscape can confer benefits on the agricultural uses. In many countries there is a need to optimize land use. The solving of questions in this field requires in-depth insight into agricultural practices, the technical possibilities for reshaping the land, by, for example, land amelioration and the impact on ecological parameters. Research into such matters is complex. It involves at least the domains of ecological economics and of land-use planning. Further, policy support for agricultural wildlife conservation requires an assessment of the trade-offs at the aggregate, regional level. At this level, the spatial element, or the 'where' question, is of special interest from both the ecological and the economic point of view.

Ecologically, it can be argued that, where the spatial distribution of landscape elements suitable for the species studied is such that adjacent land can be easily traversed, the propagation of the species has a much better chance than in a landscape where the adjacent land shows great impedance. Where nearest-neighbour fields suitable for the species are not very remote, the chances for survival are great. Where barriers, physical and/or induced by agricultural activities, exist, the chances are less.

Economically, the 'where' question is of importance because of the advantages of selective control, such as protecting most where it will be most effective and least costly. Economic research usually focuses on costs involved in wildlife management regimes and on the ecological benefits gained. For a least-cost strategy for a specific region, the economic advantages of selective control have to be taken into account. Selective control requires not only identifying the best wildlife conservation methods but also answers as to where to apply them. In this field, geographical information systems (GISs) offer important opportunities, as they provide the foundations necessary to develop spatial correlation functions, such as the dispersion–colonization process and its interactions with agricultural land use. Despite their significant potential to account for the heterogeneity of the natural environment, in practice GIS techniques are still rarely used by environmental and resource economists. This chapter presents the results of a study in which economic cost–benefit analysis was combined with a GIS-based and ecology-orientated land-use planning model.

The outline of this chapter is as follows. In the next section, farming and wildlife conservation from an ecological and regional perspective, the theoretical background of our project, are discussed. There then follow two sections – field margins, economic costs and ecological benefits, and modelling dispersion – which present the economics and landscape-planning part of the project. The research case is then presented, in which the insights from both disciplines are combined, and finally the outcomes, the methods used and the outline of plans for further research are discussed.

FARMING AND WILDLIFE CONSERVATION FROM AN ECOLOGICAL AND REGIONAL PERSPECTIVE

The spatial context of economic advantages or disadvantages of agricultural control measures is of importance when studying individual farm activities within a regional, say landscape, context, for the purpose of wildlife conservation and environmental protection. Where, for example, the conservation of natural values is at stake, the synergy of individual farm measures is relevant and needs to be identified. From an agricultural point of view, farming measures are usually ordered according to the two main dimensions of agricultural production that interact with environmental considerations: the extent of area employed in production and the intensity of input use on the area. This is commonly referred to as decisions regarding the extensive margin and intensive margin, respectively (Antle and McGuckin, 1993). A decision at the farm level as to what crops to grow determines which plants will be where and thus influences field-level processes, implying interrelations between the two dimensions. Similarly, but from an ecological point of view, ecological processes occur at levels ranging from the deoxyribonucleic acid (DNA) level of plants and animals to the (plant and animal) unit, community and landscape level.

It is to be understood that the propagation of species is not primarily a matter to be dealt with at the field or farm level but much more a case to be studied across several farms, that is at a landscape level. Basically two landscape variables for successful growth are to be taken into consideration: (i) the quality situation, referring most often to quality of land (water, air) in terms of availability of suitable land; and (ii) the possibility of access to the target site (van Dorp, 1996). When speaking of the quality situation, most often reference is made to habitat quality for selected species. Land-use measures affect ecosystem variables, such as the pH, the moisture content of the soil, the availability of macronutrients (nitrogen (N), phosphorus (P) and potassium (K)), which influence the biological preferences of the species studied and thus the abundance of the species. References concerning the accessibility of target sites reflect the need for and the drive of species to (re)colonize areas. Distances and impedances induced by the size and quality of the land influence the actual colonization. Spatial conditions between relevant habitats and 'travel quality' in gap areas are landscape variables to be studied and to be taken into consideration.

FIELD MARGINS, ECONOMIC COSTS AND ECOLOGICAL BENEFITS

Three types of wildlife conservation measures in agriculture can be distinguished: (i) along the field (that is, margins); (ii) within the field; and (iii) in between two crops in the rotation (fallow land, stubble field). In this study, we focus only on the first category. In arable fields, the largest number of plant species is found in the outer few metres of the crop. Crop edges are also more attractive for fauna than the field centre (de Snoo, 1995, and references therein). At the same time, unsprayed field margins are of special interest for Dutch agriculture, given the environmental policy for pesticide concentration in surface water. Large quantities of pesticides are used along the edge of fields; not only is the normal pesticide use on the crop involved, but also specific spraying of ditch banks and field edges. In economic terms, crop edges are less valuable than the field interior. Management of the edges often requires additional effort – in the case of wedge-shaped fields, for instance – whereas yields from the edges are often lower. Summarizing, it can be said that field-margin management offers special opportunities to integrate economic, ecological and environmental aspects.

Implementing unsprayed field margins can be done in several ways. Fallow field margins or grass strips are examples of changes concerning the extensive margin, whereas unsprayed crop strips imply changes regarding the intensive margin. In this study, 3-metre-wide unsprayed crop strips[1] were considered for winter wheat (WW), ware potato (POT) and sugar beet (SB), the three most important arable crops in The Netherlands. Costs were assessed by partial budgeting, by comparing expenditure on pesticides and the yield level with the situation in sprayed crop margins:

$$Yc_j = (p_j \times q_j) + SUBS_j \quad (1)$$
$$Ym_j = [(1 - a_j)\, p_j \times q_j] + SUBS_j \quad (2)$$
$$Yum_j = [(1 - a_j)\,(1 - b_j)\, p_j \times q_j] + SUBS_j \quad (3)$$
$$GMm_j = Ym_j - VC_j \quad (4)$$
$$GMum_j = Yum_j - VC_j + Sav_j \quad (5)$$
$$C_j = GMm_j - GMum_j \quad (6)$$

where

Yc_j = gross return of crop j in the field interior (Dutch guilders (NLG) ha^{-1});[2]
Ym_j, Yum_j = gross return in the sprayed and unsprayed margin of crop j (NLG ha^{-1});
p_j = price output of crop j (NLG kg^{-1});
q_j = yield of crop j in the field interior (kg ha^{-1});
$SUBS_j$ = subsidy of crop j if applicable (NLG ha^{-1});
GMm_j = gross margin in the sprayed field margin of crop j (NLG ha^{-1});
$GMum_j$ = gross margin in the unsprayed field margin of crop j (NLG ha^{-1});
a_j = yield reduction of crop j margin versus field interior;
b_j = yield reduction of crop j without versus with pesticide use;
VC_j = variable costs of crop j (NLG ha^{-1});
Sav_j = savings on pesticides for unsprayed margin of crop j (NLG ha^{-1});
C_j = costs of an unsprayed margin of crop j (NLG ha^{-1}).

To assess the ecological benefits of unsprayed field margins, a yardstick[3] for biodiversity on farms, developed by the Dutch Centre for Agriculture and Environment, was used in the present study. The score on the yardstick is the product of the number of units resulting from a census and a rating score (1–100 'points'). This rating score is based on the ecological importance of the species as determined on the basis of rarity, trends in size of the population and international importance (Buys, 1995). Non-spraying particularly propagates the biodiversity of dicotyledonous species and flower-visiting insects. Being an important bioindicator for wildlife quality, the occurrence of butterflies within the area was taken as indicative for total biodiversity.[4] Other species groups that make up the yardstick for biodiversity are not used in this study.

$$NVm_j = S_{i=1,...I}\,(ER_i \times Nm_{ij}) \quad (7)$$
$$NVum_j = S_{i=1,...I}\,(ER_i \times Num_{ij}) \quad (8)$$
$$NV_j = NVm_j - NVum_j \quad (9)$$

where

NVm_j = wildlife value in the sprayed field margin of crop j (score km^{-1});
$NVum_j$ = wildlife value in unsprayed field margin of crop j (score km^{-1});

ER_i = ecological rating of butterfly species *i* per unit;
Nm_{ij} = number of butterfly species *i* in the margin of crop *j* (km^{-1});
Num_{ij} = number of butterfly species *i* in the unsprayed margin of crop *j* (km^{-1});
NV_j = wildlife value in the unsprayed field margin of crop *j* (score km^{-1}).

Data for eqns (1)–(9) were available from an experiment by the Centre of Environmental Science, Leiden University, on 16 arable farms during 1990–1994 in the Haarlemmermeerpolder near Amsterdam (de Snoo, 1995).

As the GIS model used was unable to distinguish more than one field-margin type per ditch, the cost–benefit analysis was applied to combinations of field margins. All field margins are located alongside a ditch, so a combination was made up of the margins in the crops on two adjacent fields. Next, the costs C_k of a field margin combination *k* and the increase in natural values NV_k were assessed (Timmerman and Vijn, 1996). Table 14.1 presents an overview. The ratios will be used in the calculation of the ecological infrastructure to link field margins presented in the next section. Positive effects on water quality as a consequence of reduced drift from pesticides to bordering ditches are not accounted for in this study.

MODELLING DISPERSION

With the ecological landscape models ECONET (Jurgens, 1992), a landscape can be modelled and measured based upon the method of colonization to be taken into consideration, and best corridor areas are identified by means of the minimum-spanning tree method. The minimum-spanning tree method links all eco-objects together in such a way that the sum of the length of the links is minimal and all eco-objects are connected.

Table 14.1. Costs and biodiversity of field-margin combinations, Haarlemmermeer.

Field margin combination	Costs (C_j) (NLG km^{-1})	Wildlife value (NV_j) (Score km^{-1})	Ratio of costs and wildlife value	Weight*
WW/WW	111.54	1868	0.06	1
WW/POT	0.77	1652	0	1
WW/SB	565.38	1293	0.44	3
POT/POT	−110.00	1436	−0.08	1
POT/SB	454.61	1077	0.42	3
SB/SB	1019.22	718	1.42	9

*The GIS model can only handle impedance values (weights) 1–9; ratios of costs and wildlife value were therefore transformed accordingly.

ECONET1 models the situation where the colonization process is airborne and the dispersal source can be pinpointed in space, ECONET2 where the ECONET1 colonization is limited to part of the land (for example, propagation of water-borne plants limited to the existing canal, river or ditch system in the area studied), ECONET3 where the colonization is airborne but constricted by obstructing areas of considerable size and form, and ECONET4 where the ECONET3 colonization is limited within a network or corridor system and where habitats are of considerable size (Jurgens, 1992, 1993, 1996). ECONET4 is especially suited to the determination of field-margin networks; given the position of (a small number of) existing unsprayed field margins, it will assess where additional ones have to be located to ensure dispersion. ECONET4 has the option of using weights for the network elements, enabling, in addition to the dispersion condition, economic preferences in the assessment of the network to be included.

THE RESEARCH CASE

In this section, we present an application of the described approach for the situation of the Haarlemmermeerpolder, with the main objective of identifying a wildlife–cost frontier, at the regional level. A wildlife–cost frontier relates the minimum losses in farming profits associated with attaining particular ecological targets. To this end, three steps were taken: (i) digitization of the research area; (ii) determination of the distribution (frequency or spatial) of crops and of field-margin management for the baseline situation; and (iii) assessment of the ecological network and of the wildlife–cost frontier for five strategies.

For the first application of the approach, a geographically simple research area was chosen, made up of 36 farms, each of 20 ha, in which all farms have an identical cropping pattern (WW–POT–WW–SB). The result in Fig. 14.1 reflects the situation in the Haarlemmermeerpolder. The location of the crops was assessed by random selection. Next, field-margin management in the basic situation had to be assessed. Since there was no empirical information about this, each of the 36 parcels was ascribed a type of field-margin management out of four options by means of random selection. From the results of the random procedure, a sequence was chosen in which sprayed margins were relatively frequently represented (Table 14.2). For this baseline situation, an ecological network can be assessed by means of ECONET4.

Five strategies were distinguished in the assessment of the ecological network and were compared with the baseline situation A:

A = baseline, random allocation
BI = assessment according to a strict ecological point of view: minimum-spanning tree, no dispersion

BII = assessment according to a weak ecological point of view: minimum-spanning tree, but with dispersion (100–600 m)
CI = BI plus economic preferences to be included
CII = BII plus economic preferences to be included

With strategy A an incoherent network is created in the research area, based on the individual choices made by the farmers. Hence, this structure is derived directly from the results of the random procedure described. The result is visualized in Fig. 14.2. Table 14.4 (below) shows that the total costs of this strategy are NLG 606 for the area of 720 ha, being the total of the costs for the different types of field margin given in Table 14.2. In the same way, the total wildlife value of strategy A was calculated.

Strategy BI implies the determination of a network connecting all the unsprayed margins, present in the baseline situation according to the minimum-spanning tree method. Such a network would enable dispersion by all organisms found in the margins. Figure 14.3 shows the printout of ECONET4 for strategy BI. The double lines indicate where additional unsprayed field margins have to be established. Table 14.3 indicates that additionally 300 m of WW/WW margin, 900 m of WW/SB, 400 m of WW/POT, 200 m SB/SB,

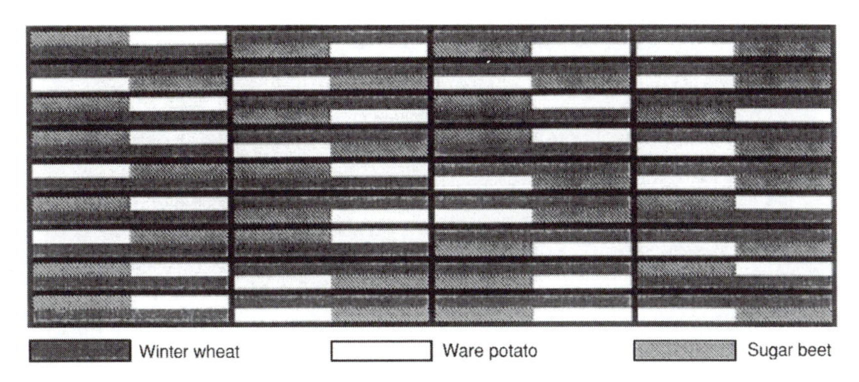

Fig. 14.1. Random cropping pattern in the research area (36 parcels of 20 ha).

Table 14.2. Costs, biodiversity and frequency (randomly assessed) of four types of field margin, Haarlemmermeer.

Type	Costs in NLG per farm	Wildlife value (score on wildlife yardstick)	Frequency (no. of farms)
1. All margins* sprayed	0	0	20
2. POT margins unsprayed	–33	430.8	3
3. WW margins unsprayed	66.94	1120.8	8
4. POT and WW unsprayed	3.94	1551.6	5

*Every margin is 3 m wide and 600 (100 + 500) m long.

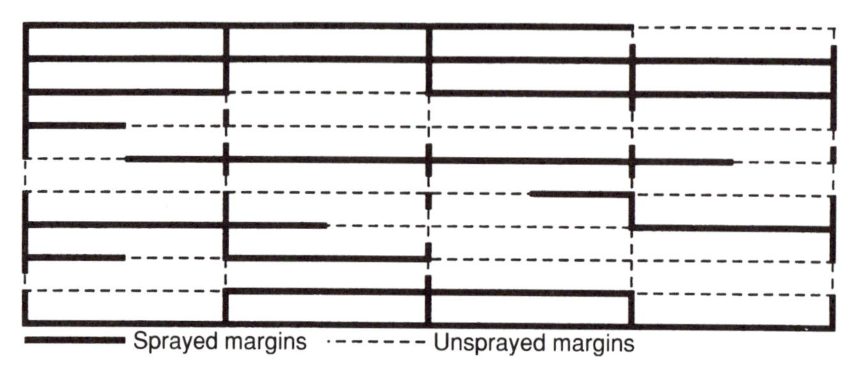

Fig. 14.2. Baseline situation (A).

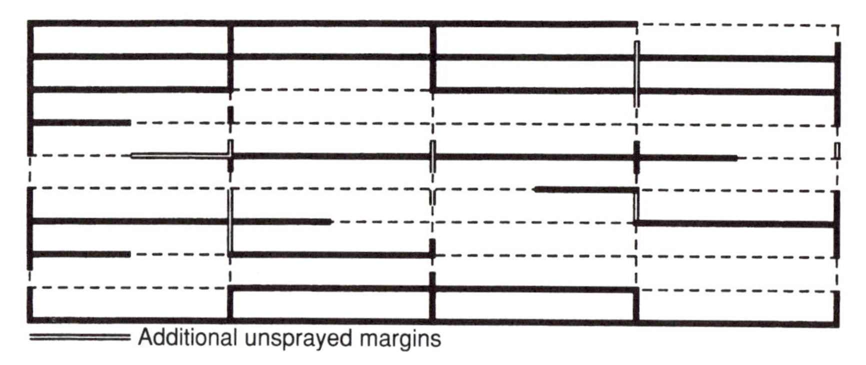

Fig. 14.3. Assessment according to the ecological point of view (BI).

100 m SB/POT and 100 m POT/POT are required. The additional costs total NLG 780 and the additional wildlife value totals 2780 points (Table 14.4, strategy BI). In the same way, the costs and wildlife values of the other strategies can be calculated. Strategy BII (100) departs from the situation where the species (butterflies in this case) can 'travel' 100 m. Less effort is required to establish a network ensuring dispersion, but at the same time less extra wildlife value is added. Under the condition of a dispersion of 100 m, the unsprayed margins need not be connected in a closed network. Results for this calculation, as well as for comparable calculations for a dispersion of 200–600 m, are given in Tables 14.3 and 14.4. ECONET4 printouts for these strategies are not shown here.

The results of strategies CI and CII indicate the advantages of selective control, combining the ecological and the economic points of view. Figure 14.4 visualizes the outcome for strategy CI, where the costs-to-wildlife ratios are used as weights in the calculation with ECONET4. Not the minimum distance but the costs per unit of wildlife for each of the optional connecting strips are decisive in the assessment of the ecological network. The costs of strategy CI

Table 14.3. Metres of extra field margin required for an ecological network under the different strategies.

Field-margin strategy	WW/ WW	WW/ SB	WW/ POT	SB/ SB	SB/ POT	POT/ POT
BI	300	900	400	200	100	100
BII (100)	200	900	400	100	100	100
BII (200)	200	300	200	100	100	100
CI	1200	300	200	100	100	100
CII (100)	1100	300	200	100	100	100
CII (200)	1100	300	200		100	100
CII (400)	1100	200	100		100	100
CII (600)	1000					

Table 14.4. Costs and wildlife value for each of the strategies for the case-study area.

Strategy	Additional costs (NLG)	Additional wildlife value (score on wildlife yardstick)
A = baseline	606	18,017
BI	+781	+2780
BII (100)	+668	+2521
BII (200)	+328	+1415
CI	+440	+3283
CII (100)	+429	+3096
CII (200)	+327	+3024
CII (400)	+270	+2730
CII (600)	+112	+1868

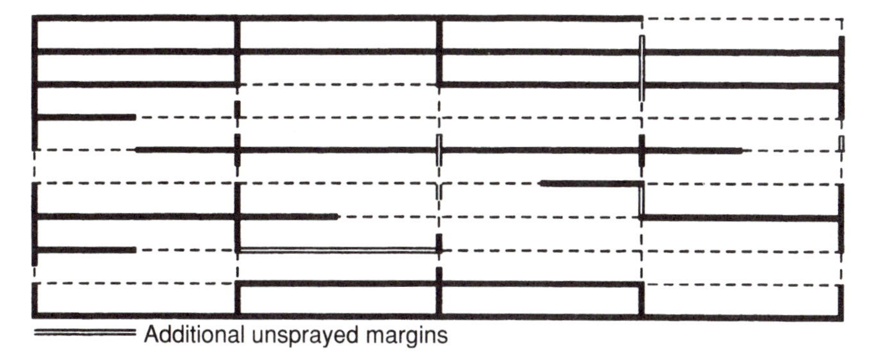

Fig. 14.4. Assessment according to the farm-economic and strict ecological point of view (CI).

compared with strategy BI are 44% less, whereas the wildlife score is 18% higher. For strategy CII (100) compared with BII (100), the differences are 36% less cost and 23% more wildlife. In the determination of the spatial pattern, field margins with sugar beet (WW/SB, POT/SB and SB/SB) are particularly avoided (see Table 14.3). This is in line with the high impedance weight for field margin combinations with sugar beet (see Table 14.1).

The results of the calculations for the different strategies, given a random-chosen baseline situation for the research area, can be summarized by means of a wildlife–cost frontier (Fig. 14.5). The same series of computations was repeated for other baseline situations. Different wildlife–cost frontiers from these situations obtained as the baseline situation, regarding both cropping pattern and unsprayed field margins already present, have a significant effect on the outcome. Results of these computations are not shown here.

DISCUSSION AND CONCLUSIONS

The wildlife–cost frontiers in Fig. 14.5 show the advantages of selective control in wildlife conservation: more biodiversity at lower costs. The modelling approach of ECONET further enables the spatial identification of ecologically desirable field-margin management, that is, which farmers should participate in a wildlife conservation programme and where the unsprayed zones should be located. The approach described is a first step towards tailor-made advice for agricultural wildlife conservation. Given the discussion of imposing a so-called

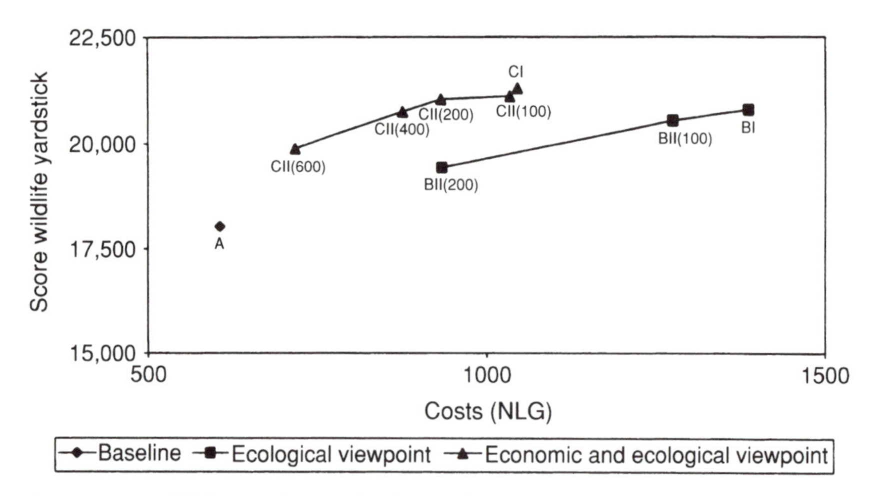

Fig. 14.5. Wildlife–cost frontier for the total area (720 ha), given random field-margin management at the farm level and different strategies (A, BI–BII, CI–CII) of establishing ecological networks.

wildlife-quality minimum (Algemene Natuur Kwaliteit) for agricultural areas in The Netherlands, information on management options available and the associated costs and wildlife scores can be very valuable to support policy development and private initiatives by groups of farmers, the so-called 'environmental' cooperatives.

In this study only field margins were considered. For a complete picture to be given, two other options need to be included: wildlife conservation in the field and in between two crops in the rotation (fallow land, stubble field). The latter addition can be realized easily by means of a sequential assessment of different rotational patterns. Accounting for wildlife conservation within the field is more complicated. It would require a grid-based GIS model to address the network question. For a more detailed farm-economic overview, linear programming preferably followed by a perception analysis among the farmers in the region, could be used instead of partial budgeting (Van der Meulen *et al.*, 1996; Wossink *et al.*, 1996). This, however, does not affect the basic methodological principle, as demonstrated in this chapter. A further elaboration would be to combine the output from the farm-economic analysis with the land-use planning model ECONET into a dynamic programming framework (Wossink and Jurgens, 1995). This would enable optimization of the strategy search for the region and would enhance the simulation procedure used in this study.

NOTES

1 In ware potato, the use of fungicides was allowed in the margins.

2 NLG 1 = 2.15 European currency units (ECU).

3 For the yardstick, 199 species of vascular plants, 17 species of mammals, 77 species of nesting birds, 14 species of wintering birds, seven species of amphibians, two species of reptiles and 26 species of butterflies were selected.

4 In this chapter, biodiversity, ecological benefits and wildlife are used interchangeably. Biodiversity includes three aspects: (i) ecosystems and natural processes; (ii) species; and (iii) genetic variety. The focus here is on the second aspect; the other two are only partly considered, namely, as far as they relate to agro-ecosystems.

15

Impacts of the European Union Reform Policy after 1996 in North-east Germany: Landscape Change and Wildlife Conservation

Hans-Peter Piorr, Michael Glemnitz, Harald Kächele, Wilfried Mirschel and Ulrich Stachow

INTRODUCTION

The European Union (EU) reform of the Common Agricultural Policy (CAP) assigns increasing importance to market mechanisms as a relevant factor for a balanced rural economy, based on price reduction and compensating the resulting income effects by provision of direct income payments and an early-retirement schedule for farmers. An obligatory set-aside scheme, with 12% fallow in the case of rotational set-aside and 17% in the case of permanent set-aside, was introduced. These measures were expected to be interlinked with benign environmental side-effects. Reduced agricultural prices lead to higher thresholds in integrated pest management systems and hence to a decreased application of pesticides. The same trend should become evident when analysing the consumption of fertilizers.

Measures aiming directly at environmental goals were Council Regulation (European Economic Community (EEC)) 2078/92 of 30 June 1992 and Council Regulation (EEC) 2085/93 of 20 July 1993, improving the protection of the environment, maintenance of the countryside and restoration of agricultural landscapes and compensating the farmers for the provision of environmental services. Council Regulation (EEC) 2092/91 of 24 June 1991, regulating the labelling of organic products, aimed for a standardized concept of organic products throughout the EU. The goal was consumer protection and, by introducing unified standards, to improve the prospects for the marketing of organic products. This might then also have positive effects on surplus reduction and environmental benefits.

 The Economics of Landscape and Wildlife Conservation (eds S. Dabbert, A. Dubgaard, L. Slangen and M. Whitby)

This chapter is especially dedicated to the change of the character of agricultural landscapes and the effects of a reduced diversity of land-use systems and cropping systems on wildlife conservation after the introduction of the CAP reform.

METHODS

Economic Model

Economic effects of the EU reform were estimated on the basis of a normative–empirical model. Static frame conditions were assumed for the situation after 1996. The 'after EU-reform' period is described as a balanced situation. No dynamic processes during the time of adaptation are assumed. The economic model is an activity-based simulation model, including parts of a programme-planning approach (Steinhauser *et al.*, 1992). One hundred and five farm models were developed. Each farm represents several districts according to the former Agricultural Cooperative for Plant Production (LPG-P) of the German Democratic Republic (GDR) (AdL, 1989). To determine the capacities of the regional farms, the arable land and grassland of the corresponding districts are totalled. The actual quotas for milk and sugar beet were determined by the extent of production before 1989 (Bundesregierung, 1992). Housing requirements were estimated according to the volume of livestock production in 1992. As livestock housing in eastern Germany is generally in a bad condition, it is assumed that further use requires repair costs amounting to approximately half the costs of new buildings (Kächele, 1993). The requirement for labour was defined as an endogenous variable for each regional farm.

Six standard production methods for arable land, two standard methods for grassland and six standard methods for animal husbandry were formulated. State-of-the-art technology was assumed. Yields and fertilizer application for plant production were distinguished according to the spatial distribution of soil quality and hydrological conditions, known from the databases GEMDAT (database for spatial land-use distribution in the former GDR) and DAWIP (database for climate and weather in the former GDR) (Lieberoth *et al.*, 1977; Bork *et al.*, 1995).

For estimating the long-term effects of the EU agricultural policy, the land rent instead of the gross margin was chosen as the decision criterion as to whether farmland is cultivated or left fallow. The overhead costs were calculated on normative data, according to standard values published by KTBL (1993). Standard data do not depict the special situation of the farms in eastern Germany a few years after the unification. For these farms, overhead costs include, for example, costs for credits that were given before 1989 for investment without any use for the actual production process. Until now, there are no final regulations to rectify this problem. As there were no reliable empirical data available, these input data are afflicted with considerable uncertainty. The scenario technique offers the possibility of handling this uncertainty by formulating different scenarios for the upper and the lower boundaries for uncertain parameters. In case of favourable regulations, overhead costs of 200 Deutschmark (DM) ha^{-1} (scenario I) are assumed; in case of unfavourable regulations, overhead costs are assumed to come to 400 DM ha^{-1} (scenario II).

- Scenario I: high degree of adaptability of farms to the new economic situation.
- Scenario II: low degree of adaptability of farms to the new economic situation.

Results of the economic model were the spatial structure of plant production systems, subsidized set-aside acreage and fallow acreage at the district level, as well as the spatial structure of animal husbandry and the extent of agricultural employment at the level of regional farms.

- Subsidized set-aside acreage: arable land, subsidized if no longer cultivated for food or fodder production, permanent or rotation fallow, according to national regulations.
- Fallow acreage: abandoned arable land and pasture disposable for alternative land use.

Alternative Land Uses of Abandoned Farmland

The results of the economic model gave the acreage of farmland that will be taken out of production after the EU reform is put into action in 1996. Subscenarios were created for alternative uses of abandoned farmland (Mirschel and Glemnitz, 1995):

- Subscenario a: afforestation of acreage taken out of production.
- Subscenario b: establishing grassland on abandoned arable land and maintaining the open character of the abandoned grassland (open range).

The decisions as to which acreage of abandoned arable land should be afforested with high priority (subscenario a) included different parameters: soil type, annual precipitation, mean temperature in July, days in May and September with temperatures below 0°C and status of wildlife conservation area. Taken together with the algorithms, this resulted in an acreage with pine, oak-mixed or beech-mixed afforestation. If the conditions for afforestation according to the algorithms were not fulfilled, the area was treated as a landscape conservation unit with mulch cultivation.

Subscenario b was defined as a landscape conservation scenario, with a high priority for the establishment of the environmental benefits of abandoned farmland. The goals were integrated in the algorithms listed in the following priorities: (i) control of water erosion; (ii) protection of ground and surface water from emissions; (iii) protection of species and habitats, (iv) control of wind erosion; and (v) groundwater recharge. As listed in this sequence, every rural commune was checked for the possibilities and necessities of realizing these aims. First, the erosion-sensitive areas of a commune were laid fallow if farmland was taken out of production. If more abandoned farmland was available in a commune, sensitive ground- and surface-water areas of the commune were taken out of production. In this order, the acreage of abandoned farmland was distributed following the environmental priorities of each communal district.

Land-use Classification

A change of the landscape character is to be expected if high percentages of farmland are abandoned and converted to open range or are afforested. An analysis of possible changes of the built environment was performed for each of the 509 rural communes, classifying land use according to dominant types: plough-land, grassland, forest and miscellaneous (Table 15.1).

Model of Land-use Diversity and Diversity of Cropping Systems

The model of land-use diversity integrates the number of different land-use types and the regularity of their distribution in each rural commune. The model is based upon the Shannon index:

$$HS = -1 \times \sum \frac{n}{a} \times \ln \frac{n}{a}$$

where a is the number of main land-use types and n the number of area units which cover a rural commune. An increasing number of land-use types leads

Table 15.1. Criteria for the land-use classification.

Land-use classification	Criterion
Plough-land	> 60% plough-land
Open range	> 60% grassland
Forest	> 60% forest
Plough-land–forest	33–60% plough-land, 33–60% forest and < 20% grassland
Plough-land–open range	33–60% plough-land, 33–60% grassland and < 20% forest
Open range–forest	33–60% grassland, 33–60% forest and < 20% plough-land

to higher *HS* values. The same method was applied for the analysis of the diversity of cropping systems (Stachow, 1995).

RESULTS

Landscape Change

The study area

The study area is located in the eastern parts of Brandenburg and Mecklenburg-Vorpommern, with Berlin in the south and the Baltic Sea in the north, parallel to the Polish border (Fig. 15.1). It covers 9300 km^2 and includes 13 rural districts with 550 rural communes. Data were available from 509 rural communes. Figure 15.1 illustrates the rather low productivity of the agricultural soils in this area, with an average index of land quality (maximum 100) of 35, ranging from 19 in Ueckermünde to 44 in Prenzlau. The relation between soil conditions and types of land use is characterized by a coefficient of determination of $r^2 = 0.74$, explaining higher percentages of farmland with an increase of soil productivity. Additional to the problems related to marginal soils, low precipitation, ranging from 470 to 550 mm $year^{-1}$, limits the productivity.

Results of the economic model

The economic model provided the database of the study concerning the acreage of abandoned farmland after the completion of the EU reform. Table 15.2 indicates the possible extent of landscape changes in north-east Germany. The initial situation of the study area, characterized by 493,000 ha arable land, including 138,000 ha set-aside, 95,000 ha grassland and 222,000 ha forest, is changed. On condition that farms respond to the new economic situation with a high degree of adaptability (scenario I), 26,000 ha of farmland are abandoned. Due to a low degree of adaptability of farms (scenario II), 187,000 ha of farmland are taken out of production. Including the subsidized set-aside area, 164,000 ha of farmland (scenario I) are no longer cultivated for food or fodder production. In scenario II, 286,000 ha are removed from former agricultural production.

Changes of the landscape character

According to the definitions of Glemnitz and Mirschel (1995), the situation prior to the introduction of the EU reform was dominated by plough-land and forest (Table 15.3). The reduction of arable land in scenario I is caused by an increase of subsidized set-aside farmland. Keeping the same extent of forest as prior to the agrarian reform, this land-use alteration would favour the open-range character in north-east Germany.

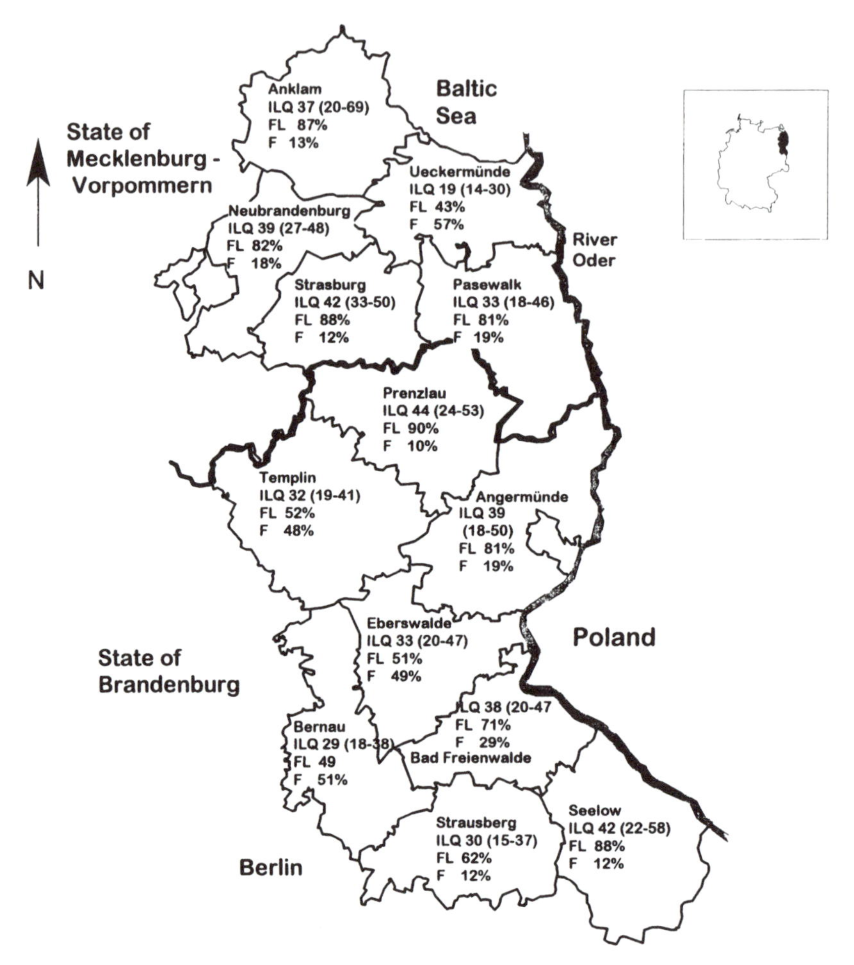

Fig. 15.1. Study area with 13 rural districts, index of land quality (ILQ), percentage farmland (FL) and percentage forest (F).

Scenario II, predominantly determined by abandoned plough-land, emphasizes the forest type if afforestation is given preference, as in scenario IIa. If the open character of landscapes is maintained (scenario IIb), a well-balanced ratio between all land-use types would become apparent. Such substantial alterations of the appearance and character of landscapes are connected with manifold socio-economic and ecological implications. The whole north-east of Germany, with the lake district and the coast of the Baltic Sea in Mecklenburg-Vorpommern and the vast number of wildlife reservation areas in Brandenburg, belongs to the green-belt recreation areas of the urban centres of Berlin, Hanover and Hamburg. A one-sided increase of one landscape type could severely disturb the character of this built environment, with its harmonious

Table 15.2. Land use (ha) in the study area after 1996 due to the new economic situation after the EU reform 1996.

Status quo situation 1992	Farmland 587,595 (plough-land 492,854, grassland 94,741), forest 221,696			
Change after 1996	Scenario I		Scenario II	
Set-aside	138,000		98,571	
Fallow	Farmland 25,681 (plough-land 4030, grassland 21,651)		Farmland 187,082 (plough-land 115,939, grassland 71,143)	
Alternative land use	Ia	Ib	IIa	IIb
(a) Afforestation	9799	1332	123,503	19,136
(b) Open range	15,882	24,349	63,579	167,937
Farmland without agricultural production	163,681		285,653	

Table 15.3. Changes of the landscape character of the whole study area according to the scenarios after the EU reform.

Scenarios	% Plough-land	% Acreage with grassland character (set-aside, grassland, fallow)	% Forest	Type of landscape (dominant land use in bold letters)
Initial situation 1992	53	19.6	27.4	**Plough-land**–forest type
Ia	43.3	28.1	28.6	**Plough-land**–open-range–forest type
Ib	43.3	29.1	27.6	**Plough-land**–open-range–forest type
IIa	34.4	22.9	42.7	**Forest**–plough-land type
IIb	34.4	35.8	29.8	**Mixed** type

succession of lakes, forests and farmland. These effects have to be examined for each rural commune, because landscape changes are perceptible on this scale. Together with agriculture's decline – more than 50% of all farmers in Germany receive more than half of their income from off-farm sources – large income losses within the tourist sector could contribute to a threat to the viability of rural areas.

Table 15.4 gives an overview of how many rural communes are afflicted with this possible development. Hardly any difference between the initial situation and scenario I are detectable. Significant shifts between landscape types of rural communes are evident in scenario II. Nearly 100 communes lose their plough-land character. Scenario IIa, with a predominant afforestation of

Table 15.4. Number of rural communes with a change of their former landscape character depending on different scenarios.

Land-use classification	Initial situation 1992	Scenario I		Scenario II	
		Ia	Ib	IIa	IIb
Plough-land type	347	349	344	257	254
Open-range type	7	5	9	5	52
Forest type	37	46	37	170	37
Plough-land–forest type	54	53	53	28	17
Plough-land–open-range type	22	15	21	8	17
Open-range–forest type	2	3	6	13	70
Mixed type	40	38	39	28	62

abandoned farmland, will contribute to the conversion of 133 communes to the forest type. Those regions characterized by a high percentage of forest possibly endanger their attraction to tourists, because of the development of monotonous wooded areas, an impression that is intensified by pine afforestation, which is preferred on sandy soils. Thirty-seven rural communes will change their plough-land–forest or open-range–forest character under the conditions of scenario IIa. The effects of scenario IIb are moderate compared with scenario IIa. Only 45 rural communes are affected by a trend to a more monotonous landscape character caused by an increase of open range.

Figure 15.2 demonstrates the initial situation, which is dominated by cropland in the east near the River Oder, in the middle of the study area in the Uckermark and in the north-western parts by the Baltic Sea. Serious landscape changes would occur following major afforestation, according to the frame conditions of scenario IIa. Figure 15.3 illustrates these impressive alterations to the built environment in north-east Germany. Afflicted regions are mainly characterized by low agricultural productivity, which results from poor soils and low precipitation levels. This context concerns ecological problems with the groundwater recharge in these regions (Dannowski, 1995; Piorr *et al.*, 1997). Afforestation, especially with pine, leads to higher evaporation, mainly caused by interception by the canopy of trees. Spreading forestry in the study area, with water shortage because of the low groundwater recharge of 116 mm $year^{-1}$, therefore conflicts with the water demands of the Berlin metropolitan region. It is especially the western parts of the study area, neighbouring Berlin, that will be covered by woodlands, following the conditions of scenario IIa (Fig. 15.3).

Changes of crop rotations and landscape character

Another evident effect of the CAP reform concerns crop rotations. The structure of the subsidy scheme leads to a strong increase of rape production in

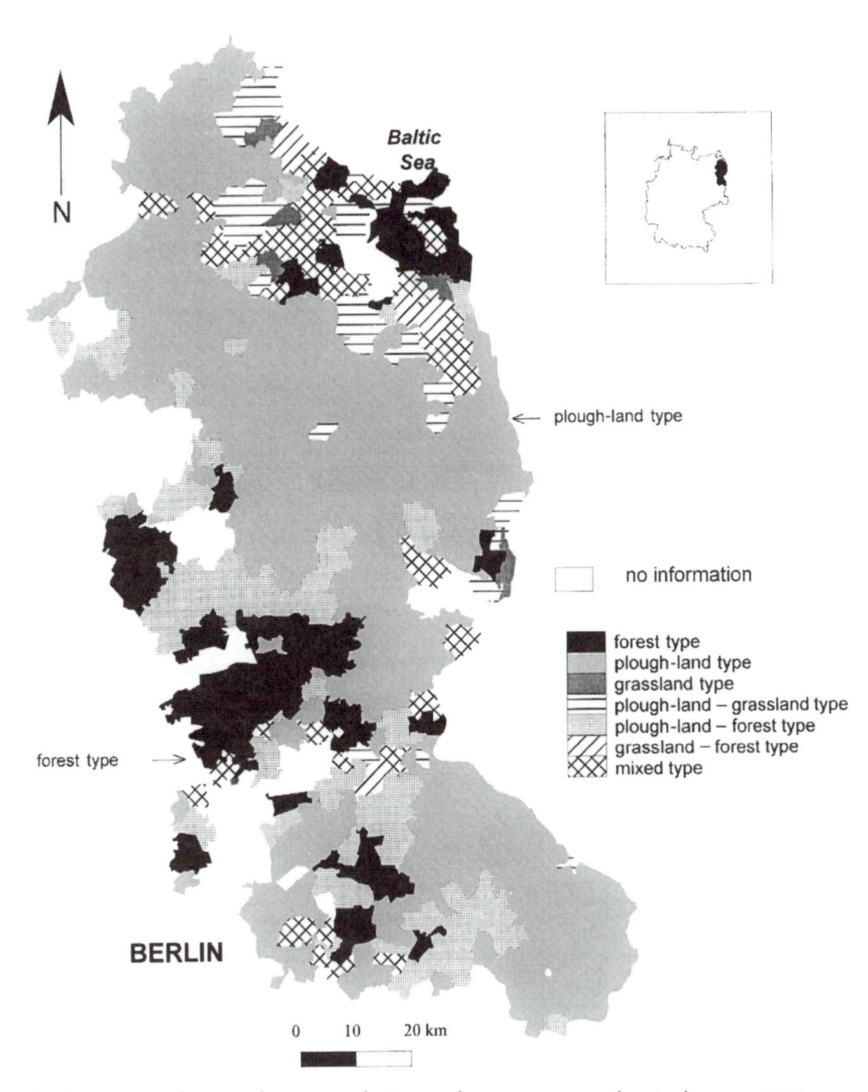

Fig. 15.2. Landscape character of 509 rural communes in the study area 1992 (initial situation).

north-eastern Germany. Judged from an agronomist's view or according to the comparative gross margin (without subsidies), the EU regulations lead to an increase of short rotations (Table 15.5). Cereal percentages have remained balanced since 1989, due to the conditions of scenario I, despite reunification and adapting to EU market regimes. A considerable reduction of the cereal cropping becomes evident in scenario II. In both scenarios, rape and maize achieve rising significance. Most profitable because of favourable subsidies,

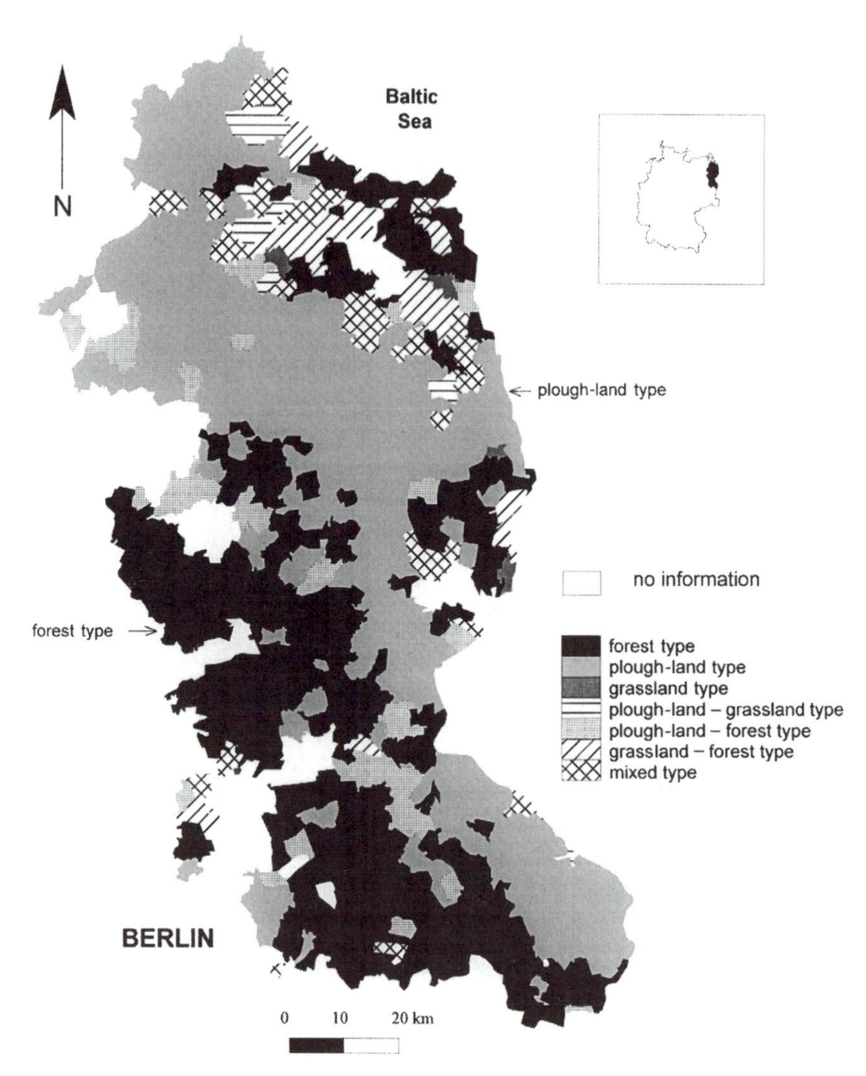

Fig. 15.3. Landscape character of 509 rural communes in the study area after 1996 due to the frame conditions of scenario IIa.

rape will cover the area that is allowed by the EU to be cropped with oil-seeds. Potatoes and alfalfa are the losers in the new organization of rotation and will possibly disappear completely.

Different side-effects of this development have to be taken into consideration. Areas that stay under cultivation will change their character after the implementation of the EU reform policy. Crop rotations will be limited to three crops (cereals, rape, maize), which will dominate on 93% (scenario I) or 89%

Table 15.5. Crops on plough-land, as percentage, in 1989, in 1992 and after 1996 due to the scenarios (without set-aside).

Crop	1989	1992	Scenario I	Scenario II
Cereals	59	58	59	47
Rape	6	13	21	26
Potatoes	8	1	0	0
Sugar beet	5	3	4	5
Maize	10	13	13	16
Alfalfa	12	12	3	5

(scenario II) of the farmland. Besides the ecological problems involved, this development will contribute to a monotonous landscape character.

Wildlife Conservation

The following section deals with the direct ecological implications of changed land use after the EU reform. Two parameters will be presented: diversity of main land-use systems and diversity of cropping systems. Both parameters are viewed as indicators for the provision of a landscape with habitats which supply the requisite conditions for numerous species. It is expected that a heterogeneous pattern with many different land-use systems is a precondition for manifold biocoenosis (Duelli, 1992). In contrast, it is supposed that homogeneous landscapes with a lower variation of habitats show a poorer species diversity.

Wildlife conservation by diversity of main land-use systems

The diversity of main land-use systems was measured as abundance and relation of arable land, open range (grassland), woodland and settlements (Shannon index). In this study, it was assumed that the 'new' land-use types, set-aside grassland and afforested land, will be ecologically similar to grassland and woodland. While no assessment of the area of settlements was made, the other land-use systems will be affected to various degrees in the scenarios. The land-use diversity will be almost unchanged in scenarios Ia and Ib (Table 15.6). This is mainly due to the fact that the total acreage of abandoned farmland is relatively small (4.4% of the farmland). Furthermore, the relation between the main land-use types will stay almost constant. In contrast, in scenarios IIa and IIb, the diversity of the main land-use systems will be profoundly affected. In these scenarios, the percentage of abandoned farmland is almost 32% of the total farmland. If afforestation prevails (scenario IIa), the diversity will decrease, on the average, because the effect of adding woodland to districts that are dominated by forests (and hence decreasing diversity) will not be compensated by the effect of adding woodland to districts where this

Table 15.6. Diversity (Shannon index) of main land-use systems (status quo situation 1992 = 100%) after 1996, aggregated on the level of 13 rural districts.

	Scenario I		Scenario II	
Rural district	a	b	a	b
Angermünde	98	99	99	107
Anklam	100	100	100	101
Bad Freienwalde	98	100	89	124
Bernau	100	100	84	113
Eberswalde	99	100	86	103
Neubrandenburg	98	100	98	100
Pasewalk	100	100	97	98
Prenzlau	101	102	106	110
Seelow	97	100	94	126
Strasburg	98	100	97	100
Strausberg	99	100	81	120
Templin	100	100	86	107
Ueckermünde	98	99	84	90
Whole study area	99	100	92	107

land-use type is not dominant (hence increasing the diversity). The decrease of diversity of main land-use systems is significant, depending on the extent of abandoned farmland ($r^2 = 0.67$).

If open range prevails (scenario IIb), the diversity of main land-use systems will increase on the average. In most districts and rural communes, most of the abandoned farmland will be converted to grassland types of ecosystems, which increase the diversity, because grassland is not abundant.

Wildlife conservation by diversity of cropping systems

Arable land is an important habitat for flora and fauna in agricultural landscapes. The diversity of agricultural cropping systems is of outstanding significance for the abundance of many organisms (Duelli, 1992; Pimentel *et al.*, 1992). Again, the Shannon index was used to analyse the diversity of cropping systems after implementation of the EU agrarian reform. As mentioned earlier, less productive soils are primarily abandoned as a consequence of the EU reform. On sites with higher soil fertility, crops will be cultivated that give high income from subsidies and/or yields. The reduction in the number of crop species leads in both scenarios to a significant decrease of the diversity. Some rural districts lose more than a quarter of their former agricultural heterogeneity (Table 15.7). The interaction between the extent of abandoned farmland and diversity of cropping systems is illustrated in Fig. 15.4. There is no doubt that especially marginal sites, which are of special importance for wildlife conservation, are endangered by this development. In east Germany many

Table 15.7. Diversity (measured by the Shannon index) of cropping systems (status quo situation 1992 = 100%) after 1996, aggregated on the level of 13 rural districts.

	Scenario I	Scenario II
Whole study area	83	84
Variance	73–95	71–96

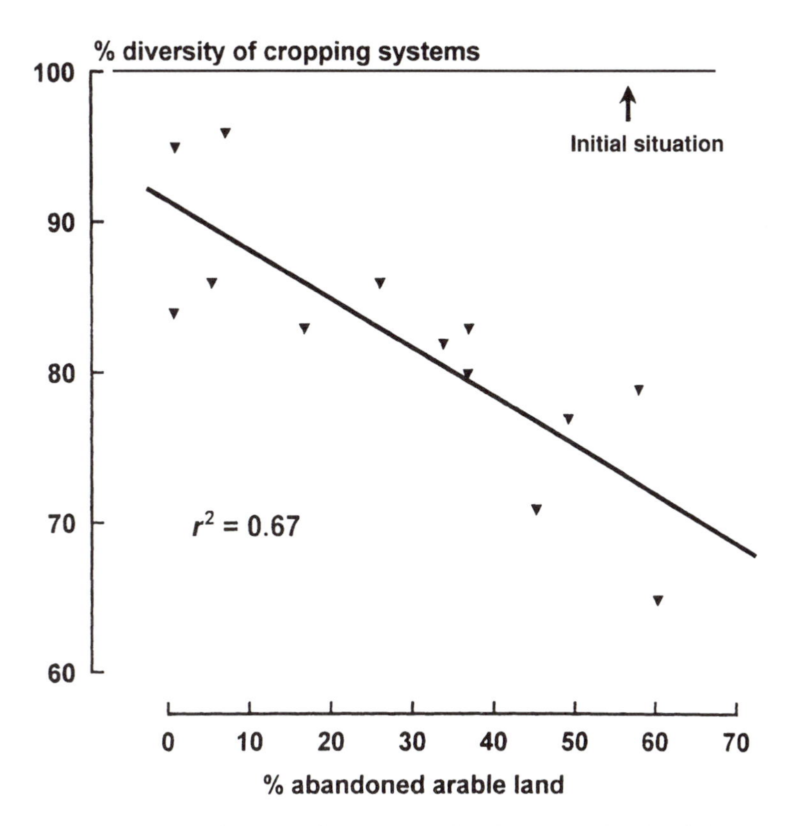

Fig. 15.4. Interactions between the extent of abandoned farmland and the diversity of cropping systems (scenario IIa).

agroecosystems on dry and sandy soils would be negatively affected by the abandonment of farmland.

CONCLUSIONS

Significant landscape changes can be expected with the completion of the EU agrarian reform. Loss of diversity in landscapes is revealed to be one important component of landscape changes. On the one hand, increasing homogeneity of landscapes is supported on arable land if subsidies for crops are given independently of site conditions. The results of this study are confirmed by developments in east Germany. For example, oil-seeds are planted on all sites, irrespective of site conditions, in order to gain maximum profits from subsidies. The overall result for both scenarios is an impoverishment of crop rotations by limitation to a few cultivated plant species. This development could be halted by stronger support of site-adapted cultivation systems, such as organic farming.

On the other hand, the risk of losing the regionally different scenery, substituted by increasing homogeneity after afforestation of large areas (scenario IIa), became apparent. To restrict the destruction of typical landscapes, regional regulations for appropriate landscape changes have to be developed parallel to the improvement of the instruments of agrarian policy.

Another characteristic of the results is the concentration of agriculture on highly productive soils and profitable crops and production systems. During recent decades, this trend has determined the development of agricultural landscapes. Obviously, this has led to the concept of segregation of areas for wildlife conservation and areas for intensive production. Our study indicates that, even if more than 30% of the arable land is taken out of production, the loss of landscape elements as habitats is not completely compensated. A contribution to the solution to this conflict is given by the introduction of low-input systems, such as organic farming. These cultivation systems are more sensitive to specific soil and climatic conditions, because of their input restrictions. Therefore they have to adapt their crop rotation and cultivation system to the site. By this means, they contribute to a manifold agricultural landscape, adapted to site-specific cropping conditions. Various authors indicate the possibility for supporting ecologically adapted landscape management on the basis of low-input farming (RSU, 1994; Weinschenck, 1994, 1995).

16

Interdisciplinary Modelling of Agri-environmental Problems: Lessons from NELUP

Andrew Moxey and Ben White

INTRODUCTION

Landscape management and conservation entail the coordinated control and modification of landscape structure, function and rate of change. This, in turn, requires knowledge and understanding of how structural components, principally the mosaic of land uses, are determined and how energy, materials and plant and animal species flow along linkages between structural components. Given that land uses are driven by socio-economic forces and that environmental linkages between different parcels of land encompass both physical and biological functional connectivity, it is apparent that a variety of academic disciplines have valid contributions to make to landscape management and conservation (Turner and Gardner, 1991; Naiman, 1992; Carlson *et al.*, 1993).

One approach to integrating knowledge and expertise from different disciplines is to engage in interdisciplinary landscape research. The jointly funded Natural Environment Research Council (NERC)–Economic and Social Research Council (ESRC) Land Use Programme (NELUP) was a research project based on such an interdisciplinary approach and ran from 1990 to 1995 at the University of Newcastle upon Tyne, UK. The project was coordinated by the Centre for Land Use and Water Resources Research (CLUWRR) and involved agricultural economists, ecologists and hydrologists working together to produce an integrated suite of quantitative models for exploring both agricultural land-use change and the consequences of land-use change within two example UK river catchments – the Tyne and the Cam. Integration between the modelling groups was achieved through a computerized decision-support system

 The Economics of Landscape and Wildlife Conservation (eds S. Dabbert, A. Dubgaard, L. Slangen and M. Whitby)

(DSS), which provided a user-friendly interface to the models and their underlying data. This chapter does not present specific empirical findings arising from NELUP, but rather offers a brief review of the rationale and structure of the project and identifies some lessons learnt during the process of conducting the research.

HISTORY AND RATIONALE

The origins of NELUP lie in discussions between several departments at the University of Newcastle upon Tyne concerning rural land use research. In particular, the Departments of Agricultural Economics and Food Marketing (AEFM), Agricultural and Environmental Sciences (AES), Civil Engineering and Geography formed the view that an approach which explicitly recognized the interactions between economics, ecology and hydrology potentially offered more fruitful insights than a single-disciplinary perspective. This led, in 1988, to the submission of a proposal to NERC and ESRC with the stated objective:

> to investigate techniques for producing a Decision Support System for land use planning comprising the socio-economic mechanisms of land allocation constrained by our scientific understanding of the physical and ecological environments. The synthesis of understanding is to be achieved mainly through the use of modelling approaches which will form the basis of the decision support system.

The research councils were sympathetic to this proposal and agreed to fund a 3-year project, starting in 1990. A 2-year extension was granted at a later date, taking the programme to the end of 1995. Given the involvement of hydrologists, river catchments were identified as the logical modelling unit and it was agreed that work would commence with the catchment of the River Tyne (an upland, mainly surface-water catchment, dominated by livestock agriculture) and then switch to the catchment of the River Cam (a lowland, mainly groundwater catchment, dominated by arable agriculture). Attention was to be focused on a general modelling framework that could be transferred to any UK river catchment. Consequently, only extant (i.e. secondary), publicly available data with national coverage were to be used. Progress was to be monitored via documentation in a 'NELUP Technical Report' series, plus biannual visits of an NERC/ESRC review committee.

ORGANIZATION

NELUP ran for 5 years, representing approximately 60 man-years of effort. This was split between four groups: agricultural economics, hydrology, ecology and DSS. The first three groups were concerned with constructing models to, respectively, explore the determinants of agricultural land-use change, explore the impacts of land-use change on water quantity and quality, and explore the impacts of land-use change on flora and fauna. The DSS group was charged with designing a 'front end' to the models. This was to integrate the models but also to provide a user interface that allowed interaction with the suite of models. The DSS group also took the lead in acquiring and storing data sets.

The economics group adopted linear programming as a methodology for constructing models of agricultural land use. Two scales were used: representative farms and the river catchment as a single (aggregate) farm. Spatial heterogeneity was represented by using a land-classification system with different production possibilities associated with each land class (Moxey *et al.*, 1995a; Oglethorpe and O'Callaghan, 1995).

The hydrology group also adopted two modelling scales. One was based around a distributed parameter model (known as SHETRAN; Système Hydrologique Européen Transport), which explicitly recognized spatial heterogeneity across a catchment by splitting the land area into 1-km-grid squares. This was very data and central processing unit (CPU) intensive. The second approach simplified this by employing a cruder degree of resolution, effectively splitting a catchment into 'lumped' areas of different land uses. This second model (known as ARNO, after the river around which the model was developed) was used as a screening tool before running SHETRAN (Adams *et al.*, 1995).

The ecology group adopted three modelling methodologies to predict the distribution of plant and animal species. The first methodology was based on simple associative models, which state the probability (derived from empirical studies) of a particular plant or animal species being found in a given (vegetative) cover type, plus the use of transition matrices, which state the probability of one cover type changing into another (that is to say, a Markov model). The second methodology was more process-based, using ecological relationships (for example, population dynamics) to construct predictive models. These first two methodologies were applicable to exploring broad-scale changes at the catchment scale. To address changes at a finer resolution, a Vegetation Environment Management Model (VEMM) was developed through ordination analysis of the British national vegetation. This VEMM allows exploration of change at a site-specific level, for example in tandem with a farm-level economic model (Oglethorpe *et al.*, 1995; Rushton *et al.*, 1995).

Although the three modelling groups were responsible for their own models, regular dialogue between groups was maintained. This was essential for two reasons. First, the groups had to share common data sets, for example land classifications. Failure to communicate here could have led to duplication of effort and, more likely, different data interpretations and treatments. Second, links between models had to be established. Failure to communicate here could have led to, for example, ecological models needing management information that was not generated by the economic model. This process of communication was facilitated by the DSS group, who, if needed, acted as go-betweens for the modelling groups.

The DSS group was also responsible for designing and constructing the DSS itself. This required collation and organization of large and diverse data sets, design and implementation of a graphical user interface (GUI) and design and implementation of on-line linkages between the suite of models provided by the modelling groups. Data storage and manipulation was conducted using geographical information systems (GIS) and relational databases. The DSS computer code is a mixture of object-orientated and modular programming and draws upon extant code libraries, such as XView. The nature of the DSS means that, to be appreciated, it has to be seen in action, rather than described with text and (static) screen snapshots. For this reason, NELUP ran a series of workshop presentations, where between 10 and 30 participants were invited to view the DSS and 'try it out'. Participants were drawn from academia, environmental agencies, local government, central government and water companies. Feedback from workshops was used, via an iterative approach based on a repetitive design cycle, to refine the DSS and the underlying models (McClean *et al.*, 1995).

The above is a necessarily brief review of NELUP. Further details are available in Volume 38 of the *Journal of Environmental Planning and Management* and in O'Callaghan (1996). Images of the DSS, along with a list of publications stemming from NELUP, may be viewed at the CLUWRR Internet World Wide Web site: http://www.cluwrr.ncl.ac.uk/. The following section of this chapter highlights some emergent issues relating to integrated land-use modelling.

EMERGENT ISSUES

NELUP was a large research project and generated a large number of academic articles. Topics addressed include: compatibility of different land-cover data sets; geographical interpolation and extrapolation of incomplete data; estimation of agricultural nitrate-pollution patterns; policy abatement of agricultural nitrate pollution; and the extent of 'halo' effects of conservation schemes (see, for example, Allanson *et al.*, 1993; Moxey *et al.*, 1995b). In addition, the DSS offers a convenient tool for conducting a relatively wide range of impromptu 'what-if' scenarios beyond those already explored (O'Callaghan, 1996). However, focusing on these research outcomes diverts attention from the research process. While the empirical outcomes are interesting, a number of equally important emergent issues revealed by the experience of conducting interdisciplinary land-use modelling also merit a public airing. Such issues are pertinent to landscape-management and nature-conservation research and yet are rarely mentioned in documentation of research outputs.

First, the formation and management of an interdisciplinary research team are not a trivial task (Anon., 1987; Topp, 1993). Although NELUP ran under the auspices of CLUWRR, it was staffed initially by academics, whose primary allegiance was to individual departments with single-disciplinary objectives, funding bodies and review processes (for example, research-assessment exercises). This led to inevitable conflict of interest: interdisciplinary work is often viewed as diluting or corrupting disciplinary purity. Overcoming this conflict took time and also careful management – including the deliberate relocation of key staff into a common building alongside other NELUP staff and away from departmental colleagues. The importance of this team management should not be understated. Economists, ecologists and hydrologists all have valid contributions to make to land-use modelling. Yet casual cross-fertilization is relatively rare, because the different disciplines effectively speak different languages and are typically encouraged to pursue single-disciplinary research. Having a common (funded) research focus and being forced to work alongside each other led to staff involved with NELUP gaining greater mutual respect for each other's disciplines and some common understanding of issues and problems arising from interdisciplinary land-use modelling work – a fact that is perhaps reflected by the frequency with which their advice is sought by research groups currently embarking on similar programmes elsewhere in the UK and abroad.

Second, data availability, compatibility and reliability are major stumbling-blocks for interdisciplinary land-use research in the UK. Data sets employed within NELUP included: June agricultural census and farm business survey data obtained from the Ministry of Agriculture, Fisheries and Food (MAFF), via the ESRC data archive at the University of Essex; soils data obtained from the Soil Survey and Land Research Centre, Silsoe; topographical and river network data obtained from the Institute of Hydrology, Wallingford; climatic data obtained from the Meteorological Office, Bracknell; land class and remote-sensed land-cover data obtained from the Institute for Terrestrial Ecology (ITE), Merlewood and Monkswood; river flow and pollutant levels obtained from the National Rivers Authority (NRA) in Newcastle, Peterborough and Bristol; and ecological survey data obtained from various local offices of English Nature, county councils and voluntary bodies. Ownership of these data is highly fragmented. Consequently, collating a database for modelling, even for only two river catchments, incurred high transaction costs, including repetitive licensing costs. Moreover, data were rarely complete, nor were different data sets readily compatible. For example, the ITE Land Cover Map of Great Britain reports land covers down to the resolution of a 25 m by 25 m pixel. It does not, however, distinguish between different types of arable crops, nor is it available as a time series. In contrast, the June agricultural census is available as a long time series, distinguishes between different types of arable crops and provides information on livestock numbers. Unfortunately, census data are available only at the resolution of a parish or groups of parishes and do not include information on management. Detailed management information can be obtained from farm surveys (for example, the farm business survey or the survey of fertilizer practice), but the spatial resolution is very crude.

Third, data restrictions and different disciplinary modelling resolutions hamper the integration of quantitative land-use models. The impetus behind spatial landscape modelling is the belief that the location of a land-use change may be as important as the type and magnitude of that change, because different site characteristics (including site history) lead to different response patterns, and that functional connectivity means that change is conditional on and conditions changes elsewhere. However, these considerations pose a challenge to quantitative modellers, since they necessitate explicit recognition of spatial variability and linkages, and yet data to adequately describe heterogeneity are typically not available. Such data-induced problems are compounded by different spatial and temporal resolutions favoured by different modelling disciplines. For example, hydrology is often concerned with catchment areas and events that occur within a short time frame (for example, storm rainfall), ecology is often concerned with specific parcels of land and a long time frame (for example, 50 years for natural-vegetation regeneration) and agricultural economics is often concerned with aspatial markets and a seasonal time frame. Attempting to overcome these problems through 'spatial articulation' and linkages dramatically increases both the conceptual and the computing complexity of models (see, for example, Boumans and Sklar, 1990; Verdiesan and Moxey, 1994).

Fourth, designing a DSS to address a wide range of 'what-if' scenarios leads to conflicts between model coverage and model accuracy. The NELUP DSS undoubtedly served as a useful focal point in drawing the three modelling disciplines together. However, because the DSS was intended to explore a wide range of 'what-if' scenarios, the underlying models had to be sufficiently broad-based to deal with a wide range of different policy instruments and market conditions. As a result, the models inevitably became rather generalist in nature. That is, they were not designed to look at any one particular problem in detail, but rather to offer general insights into a broad class of problems relating to land use. This inevitably led to some frustration for both researchers and potential users when pursuing a specific line of enquiry and suggests that a more focused modelling approach might be more desirable. Indeed, many people attending NELUP DSS workshops implied that the perceived value of a DSS lay in its ability to address a narrow range of problems in some detail. It should, however, also be noted that many potential users were wary of the DSS, viewing it as a decision-making rather than decision-support system and that some of the disquiet with the generalist approach might reflect this wariness.

Fifth, model complexity causes conceptual and validation problems. Although the DSS GUI hides the complexity of the system from users, the underlying models and data sets are rather complex. Efforts to reflect the real world led to ever-greater refinements and ever-increasing model size, causing the models to became increasingly harder to comprehend (even for the model builders) and therefore harder and more time-consuming (Brooks, 1975) to amend in response to feedback from workshops or review-committee visits. This is known as 'Bonini's paradox' (Dutton and Starbuck, 1971): making models more like the real world means that they may be no easier to understand than the real world, and this has been commented on by researchers on other large modelling projects, such as CARD (named after the Centre for Agricultural and Rural Development) at Iowa, USA. This poses the question of whether more transparent models would be preferable, even if this means that they might be less powerful (Wallace, 1994). Increasing model complexity also makes error detection and correction more difficult. While it is recognized that underlying data include errors, that the individual models include errors and that interactions between models will generate errors, it is difficult to envisage or measure the compounding pattern of these errors. Given the desire to use the models for forecasting, model validation is clearly desirable – a situation that has prompted a search for alternative validation and error-reporting techniques (Jakeman *et al.*, 1995).

SOME CONCLUSIONS

Landscape management and nature conservation entail the coordinated control and modification of landscape structure, function and rate of change. This suggests the need for an interdisciplinary approach that recognizes the influence of socio-economic factors on landscape components, principally the mosaic of land uses, and the environmental linkages and flows, principally energy, materials and species, between these components. Today, now that the issue of sustainability is established as part of landscape planning and management, it is easy to forget that, prior to the 1990s, relatively little integrated-modelling work had been conducted to explore the landscape impacts of agricultural and environmental policy measures within the UK. Indeed, NELUP owed its existence to academics at Newcastle spotting an opportunity for quantitative land-use modelling and to the willingness of NERC and ESRC to sponsor overtly interdisciplinary research at a time when the need for such work was not generally recognized. To some extent, notwithstanding the emergent issues identified earlier, the process and outcomes of NELUP research demonstrate the validity of quantitative modelling by illustrating the potential for integrating different disciplines in a manner that yields useful insights into the causes of land-use change and its impacts on landscape components and functions. The focus on river catchments was appropriate, since catchments are, or should be, functional units for agencies charged with landscape management and conservation (Naiman, 1992). To some extent, this is reflected in the rise of catchment-management plans in the UK. Indeed, as concern over agricultural water pollution, ecological damage and (non-agricultural) rural development has risen, increasing attention has been paid in general to the need for environmental awareness and the need for management strategies to be implemented at an appropriate (natural) scale, rather than piecemeal across different administrative units. In this context, the efforts of NELUP may have contributed to the wider acknowledgement of environmental connectivity and the potential for environmental knock-on effects from changes in agricultural land use induced by market or policy changes. Certainly, the NELUP experience with data collation and modelling is in accordance with calls from landscape ecologists (and others) for an integrated approach to planning and management, including the 'development and maintenance of shared information on environmental stock' (Department of the Environment, 1993).

17

An Integrated Approach to Agricultural and Environmental Policies: a Case-study of the Spanish Cereal Sector

José Sumpsi, Eva Iglesias and Alberto Garrido

INTRODUCTION

The agri-environmental Council Regulation (European Economic Community (EEC)) 2078/92 has opened up new opportunities to integrate environmental objectives within the Common Agricultural Policy (CAP) and the European agricultural sector. The agri-environmental programmes that have been developed throughout the European Union (EU) currently follow an incentive scheme. Such instruments provide farmers with environmental payments, which compensate for the extra costs of undertaking a given set of environmental practices or for the loss of profits derived from certain restrictions on their farming practices.

The implementation of any economic instrument needs to overcome several difficulties to provide the benefits the application intends to produce. For one thing, doubts will always exist about the relationship between the actions that farmers are required to perform and the benefits for the environment that farmers' behaviour theoretically produces. Secondly, the administration is hindered by a problem of asymmetric information regarding the true costs incurred by farmers to perform the tasks that provide environmental benefits. But, even if the first relationship could be scientifically established for all environmental contexts in which eligible farmers operate and the administration could evaluate the true costs of farmers with moderate accuracy, which is a big if, it remains to be ascertained whether the selected scheme provides farmers with enough incentives to 'honour' the contract they sign with the administration. This last issue is of remarkable importance, since the monitoring and administrative costs that such programmes require may represent a

 The Economics of Landscape and Wildlife Conservation (eds S. Dabbert, A. Dubgaard, L. Slangen and M. Whitby)

very significant portion of the programmes' total costs. In fact, empirical studies have demonstrated that, in some cases, monitoring costs represent about 30% of the total costs of the environmental programme (see Whitby *et al.*, Chapter 8, this volume). Therefore, we view the analysis of the nature of different schemes as a central issue in the assessment of the probability of success of agri- environmental programmes.

This chapter attempts to establish some general conclusions about which policy instruments are more appropriate for introducing environmental considerations into European agricultural policy. It also aims to provide empirical evidence of the theoretical results' robustness, using the implementation of the Council Regulation (EEC) 2078/92 on Spanish cereal lands as an illustrative example.

In order to analyse different policy instruments, we deem it most appropriate to frame the problem within the principal–agent (PA) model. The PA theoretical tenets pertain to the problem of an employer (principal) who seeks to induce a worker (agent) to perform a task whose result depends on the worker's effort which the principal can only observe through costly monitoring (Dye, 1986; Mas-Colell *et al.*, 1995).

Before defining the set of instruments to be analysed, it should be stressed that, given the remarkable difficulties of valuing the environmental benefits of these programmes, it will be assumed that the environmental payments that some of these instruments incorporate are based on the principle of cost compensation.

A large body of literature has been published on policy instruments to integrate environmental externalities with European farming activities (see, for example, Baldock, 1992; Crabtree and Chalmers, 1994; van der Weijden and Timmerman, 1994; Scheele, 1996; Sumpsi, 1996). However, the analysis will be limited to those instruments that seem to offer better chances of integrating environmental objectives within the agricultural-policy context. In particular, the following policy options will be dealt with: command and control schemes, cross-compliance schemes incentive schemes and incentive penalty schemes.

Command and Control Schemes

These schemes constitute a traditional form of regulation that involves certain restrictions on farming activities. If the established requirements are not satisfied, the farmer will be considered a lawbreaker and will be subject to a legal penalty. Several legislative bodies, at both EU and national level, include environmental restrictions to farming practices, such as habitat legislation, land-use planning legislation and forest legislation, among others. While in general the type of restrictions imposed on farming activities is soft, the Code of

Best Farming Practices establishes quite stringent requirements, which, according to the Nitrates Directive, will be compulsory within vulnerable areas.

Cross-compliance Schemes

These schemes imply the attachment of environmental conditions to agricultural support policies. We shall consider two forms of cross-compliance, which have been broadly discussed (Baldock and Mitchell, 1995). The first, denoted hereafter as cross-compliance, constitutes the conventional and most stringent form of cross-compliance. It links agricultural support payments to compliance with a given set of environmental practices. Farmers eligible for agricultural support payments are obliged to satisfy certain environmental criteria without receiving any additional compensation. The failure to comply with such environmental requirements is penalized with the loss of the agricultural support payments.

A rather softer form of cross-compliance is cross-compliance with compensation, under which farmers eligible for agricultural support payments are also obliged to comply with certain environmental requirements. However, it differs from the conventional cross-compliance in that farmers receive an additional payment to compensate extra costs incurred when undertaking the required agri-environmental practices.

Incentive Schemes

Incentive schemes are becoming widely used across Europe and have been instituted within CAP in the form of the agri-environmental programme, Council Regulation (EEC) 2078/92. Under this approach, farmers can enter the agri-environmental programme voluntarily and receive a specific environmental payment. Failure to carry out the established environmental practices implies the loss of environmental payment (Council Regulation (EEC) 746/96).

Incentive–Penalty Scheme

Lastly, we propose another policy instrument, denoted as the incentive–penalty scheme, under which farmers voluntarily enrolling on a given agri-environmental programme are entitled to an environmental payment. But, unlike the incentive scheme, the incentive–penalty scheme establishes that, once the farmer has decided to enter the agri-environmental programme, failure to carry out the environmental requirements implies not only the loss of agri-environmental payments, but also the loss of support payments (CAP reform compensatory payments).

Table 17.1. Definition of policy instruments.

Instrument	Character	Payment	Penalty
Command and control	Compulsory	None	Legal fines
Cross-compliance	Linked to compensatory payments	None	Loss of compensatory payments
Cross-compliance with compensation	Linked to compensatory payments	Environmental payment	Loss of compensatory and environmental payments
Incentive scheme	Voluntary	Environmental payment	Loss of environmental payment
Incentive–penalty scheme	Voluntary	Environmental payment	Loss of environmental and compensatory payments

Table 17.1 summarizes the basic aspects characterizing the different policy instruments that are included in the analysis.

The remaining sections of the chapter are organized as follows. The next section covers the background of the specific characteristics of the Spanish cereal sector, followed by an introduction to the theoretical framework, showing how a slight modification of the PA theory provides interesting insights about the nature of agri-environmental contracts. In the next section, an empirical application of the theoretical model to the Spanish cereal sector is developed. In the final section, we lay down some conclusions, formulate several policy recommendations based on the results and suggest further lines of research.

BACKGROUND TO THE SPANISH CEREAL SECTOR

The Spanish cereal sector differs from those of other EU countries in that a large part of the cereal production is obtained in semiarid regions under dryland regimes. The traditional production system in such areas is based on 2- or 3-year crop rotations, in which land is left idle for 1 year – hereafter termed traditional white fallow (WF) – in order to accumulate enough soil moisture and secure a higher yield in the crop immediately after the WF. The percentage of traditional WF depends on the soil quality and degree of aridity, and represents nearly 4 million ha. Well grounded within the environmental conditions, it is an extensive production system profoundly adapted to the limited natural-resource endowment.

However, when the cereal Common Market Organization (CMO) reform was introduced, it almost provoked a severe productive and environmental problem in Spain. This potential threat resulted from the exclusion of the traditional WF from the area eligible for compensatory payments, which amounts to 7 million ha of cultivated cereal land. During the first few months after the reform was put into effect, the amount of land resulting from farmers' applications for first-year compensatory payment increased to dangerous levels, with a risk of exceeding the regional reference surface and triggering the penalties scheme. Farmers, undoubtedly, were induced by the compensatory payments to cultivate cereals on all available land, including the plots that farmers would have otherwise devoted to WF (Sumpsi, 1994). Thus, the exclusion of traditional WF from the compensatory payments had the unintended effect of artificially intensifying the traditional extensive cereal-production systems in Spain. Needless to say, this was exactly contrary to the spirit of the CAP reform.

This undesired side-effect of the cereal CMO reform forced the Spanish Ministry of Agriculture to pass a regulation establishing for each region a WF index, which all cereal farmers had to satisfy. Each region's index is equal to the ratio of the 1989–1991 average land devoted to traditional fallow over the land devoted to annual crops. Through this measure, the Spanish administration managed to maintain the traditional WF, although farmers are not entitled to compensatory payments on the land required to be left idle in order to comply with the traditional WF index. The WF area established by the Spanish Order happens to be independent of the set-aside requirements established by Council Regulation (EEC) 1765/92, for which farmers do receive compensatory payments.

Lastly, in 1995, the Spanish government passed a decree that implements the Council Regulation (EEC) 2078/92 at national level. The most important measure is related to the maintenance of WF under certain environmental practices (green fallow), mainly directed towards the fight against erosion, and for the prevention of forest fires.

To summarize, there are two types of non-cultivated land in Spain: mandatory set-aside imposed by the CAP reform and the traditional WF. Our analysis is centred on the policy instruments that could induce the voluntary conversion of the traditional WF into a green fallow, conforming with the Council Regulation (EEC) 2078/92, as developed by the Spanish government.

THE THEORY OF CONTRACTS APPLIED TO THE ANALYSIS OF POLICY INSTRUMENTS

This section deals with how the analysis of monitoring costs can be framed within the PA theory. Irrespective of the type of policy option to be analysed, the administration can be thought of as the principal and farmers as the agents. The principal cannot monitor all signing farmers to check whether they comply with the set of tasks specified in the contract. Therefore, the principal is forced to rely on the threat of inspecting the agent and, if fraud is detected, impose a penalty and throw the farmer off the programme. In this sense, the classical presentation of the PA somewhat differs from ours, in that here the stochastic nature of the contract is due to the probability of being inspected, while in the conventional PA it results from the uncertainty that a set of market events will take place, with probabilities related to the effort put in by the agent.

Richard and Trommetter (1994) have also proposed the analysis of agri-environmental instruments within the framework of the PA model. Their paper focuses on the uncertainty the principal faces with respect to the number of farmers that sign the contract and on the dynamic and irreversible nature of the farmers' decision about signing environmental contracts. While they disregard the possibility of farmers putting in low effort – such as not performing the environmental practices they are supposed to carry out – and collecting the environmental subsidy that supposedly compensates the farmer for producing the service, we consider this as a central issue in the analysis of environmental instruments.

The Theoretical Model

Let farmer k be a risk-averse agent who has the option to enter an agri-environmental programme that imposes some conditions on the way he/she manages his/her fallow land and receives in return a subsidy for performing these actions. His/her utility function is given by:

$$u(Y) = E(Y) - \beta\sigma_Y \tag{1}$$

where $E(Y)$ is the expected value of income Y, β is the risk-aversion coefficient and σ_Y is the average deviation of Y.

It is assumed that the principal has complete information about the farmer's costs. He/she seeks to induce the farmer to sign a contract and perform certain environmental practices on the WF. Although a certain degree of heterogeneity among farmers should be considered, for the moment it will be assumed that the principal faces K identical farmers. This assumption will allow the analysis to be centred on the monitoring costs associated with each policy instrument. Later on, the implications of this assumption will be discussed to consider the existence of a K^* optimum and examine the degree of flexibility that each policy instrument permits.

Given that the principal has complete information about the farmer's costs, he/she can offer the farmer a contract that ensures that he/she will enter the programme and comply with its terms. Following the general assumption of considering the principal as risk-neutral, such a contract will result from the following optimization problem:

$$\min_{\gamma} [P_E + C_m\gamma] \tag{2}$$

subject to:

$$E(I + \pi_P^C) \geq E(\Gamma + \pi_{NP}) \tag{3}$$

$$E(I + \pi_P^C) \geq (1 - \gamma)E(I + \pi_P^{NC}) + \gamma E(I + \pi_P^{NC} - P) - \beta\delta(\gamma, P_E, P) \tag{4}$$

where:

P_E = environmental payment;
C_m = monitoring cost;
γ = probability of inspection;
I = farm income in cultivated and set-aside land with environmental programme;
Γ = farm income in cultivated and set-aside land without environmental programme;
$\pi_P^C = P_E - C_E$ = net incomes in WF when the farmer enters the environmental programme;
$\pi_{NP} = -C_T$ = net incomes in WF when the farmer does not enter the programme;
$\pi_P^{NC} = P_E - C_T$ = net incomes in WF when the farmer enters the environmental programme but fails to comply with the given set of environmental practices, with C_E = cost of environmental practices and C_T = cost of traditional practices, with $C_E > C_T$, since environmental practices are more stringent than traditional practices;
P = penalty for being caught committing fraud;
$\delta(.)$ = average deviation of income he/she might receive if he/she decides to commit fraud.

Equation (3) is the rationality constraint and states that the income achievable when signing the environmental contract is greater than the regular income when the farmer does not accept the contract. Equation (4) is the incentive constraint; it ensures that the farmer will comply with the requirements laid down in the contract by establishing that the expected income of complying with the environmental programme is greater than the expected income of committing fraud. The latter will be given by the expected result of the 'lottery' of being or not being caught committing fraud. The risk-aversion coefficient times the average deviation reflects the risk-averse behaviour of farmers.

With the general model sketched above, we are now equipped to specify the details that explicitly define the payment schedule of each policy instrument:

- Cross-compliance (X):

  $I = W + CP; \Gamma = W; P_E = 0; P = CP$

- Cross-compliance with compensation (XC):

  $I = W + CP; \Gamma = W; P_E = C_E - C_T; P = CP + P_E$

- Incentive–penalty scheme (IP):

  $I = \Gamma = W + CP; P_E = C_E - C_T; P = CP + P_E$

- Incentive scheme (I):

  $I = \Gamma = W + CP; P_E = C_E - C_T; P = P_E$

- Command and control (CC):

  $I = W + CP; \Gamma = W + CP - P; P_E = 0; P = P_L$

where W = net market returns; CP = compensatory payments.

Command-and-control schemes are, by definition, enforced by law. Since it is not optional but obligatory for the farmer to participate in the programme, the rationality constraint is not applicable within contracts implemented under this policy instrument. For the rest of the policy options and given the above specified parameters, it is straightforward to prove that the rationality constraint (eqn 3) is always fulfilled.

Given the payment schedule, since the cost function is linear in γ, each policy optimum is related to the minimum γ that fulfils the incentive constraint. Thus, solving the PA model expressed by eqns (2) and (4) yields a γ^* that minimizes the cost to the principal of implementing a given environmental contract and makes constraint (4) binding. We hereby limit ourselves to writing the general equation, which expresses the optimum value of the choice variable.[1]

$$\gamma^* = \frac{1+2\beta}{4\beta} - \sqrt{\left(\frac{1+2\beta}{4\beta}\right)^2 - \frac{(C_E - C_T)}{2\beta P}} \qquad (5)$$

Some general results follow from this equation:

- As expected, the probability of inspection is inversely related to the risk-averse attitude of farmers: $\gamma = \gamma^* \, (\beta, (C_E - C_T/P))$, with $\partial\gamma^*/\partial\beta < 0$ and $\partial^2\gamma^*/\partial\beta^2 < 0$.
- The probability of inspection is related to the ratio between extra costs and the penalty $(C_E - C_T)/P$.
- There is a clear trade-off between the optimum level of inspection and the penalty defined by each policy option, with $\partial\gamma^*/\partial P < 0$ and $\partial^2\gamma^*/\partial P^2 > 0$.

Given the different penalties established in each policy instrument, we can sort the γ terms and establish a comparison among the monitoring costs that will arise under each policy scheme. Given that:[2]

$$\frac{C_E - C_T}{CP + P} < \frac{C_E - C_T}{CP} < \frac{C_E - C_T}{P_L} < \frac{C_E - C_T}{P_E}$$

it becomes clear that the γ terms can be ordered as follows:

$$\gamma^*_{IP} = \gamma^*_{XC} < \gamma^*_X < \gamma^*_{CC} < \gamma^*_I$$

Lastly, since the monitoring cost is defined by $MC = \gamma C_m K$, it can be established that the total monitoring costs incurred in each policy option can also be ordered as:

$$MC_{IP} = MC_{XC} < MC_X < MC_{CC} < MC_I$$

In order to grasp the differences among the different policy options, the results written above are summarized in Table 17.2.

EMPIRICAL RESULTS

The most important agri-environmental programme developed in Spain under Council Regulation (EEC) 2078/92 is the extensification of cereal lands, which establishes a set of environmental practices that should be carried out in the traditional white fallow. Here we shall particularize the results obtained in the theoretical model to the cereal extensification programme and compare its current implementation through an incentive scheme with an alternative policy option. The results obtained in our theoretical model demonstrate that the incentive–penalty scheme and the cross-compliance with compensation require the least monitoring effort.

Moreover, according to other relevant criteria, such as acceptability and redefinition of property rights, we deem that command and control as well as cross-compliance are not adequate policy instruments when dealing with agri-environmental measures demanding stringent requirements that redefine farmers' property rights. Indeed, most of the '2078' agri-environmental programmes, and particularly the extensification measure in Spanish cereal lands, seek to generate positive externalities in rural areas. In such cases, the provision of environmental goods without financial compensation would not be readily accepted by farmers.

With regard to flexibility and efficiency criteria, Richard and Trommetter (1994) contend that there is a socially optimum number of farmers K^* that results from equating the marginal costs and benefits for each programme. In this aspect, a cross-compliance with compensation approach would incur excessively high financial costs, since all the farmers eligible for compensatory payment would, by definition, be entitled to environmental payment. Furthermore, problems of inefficient assignment would also occur, since farmers qualifying for the same environmental payment might face different costs (Crabtree and Chalmers, 1994). Under an incentive–penalty scheme, this problem could be solved through a more efficient system. For instance, an auction system

Table 17.2. Matrix results of PA model.

Ranking of MC from smaller to higher	Policy options	Payment schedule		Probability of inspection
		Environmental payment	Penalty	
1	Incentive–penalty scheme	$P_E = C_E - C_T$	$P = CP + P_E$	$\gamma_{IP} = \gamma^*_{IP}(\beta(c_E - c_T)/(CP + P_E))$
1	Cross-compliance with compensation	$P_E = C_E - C_T$	$P = CP + P_E$	$\gamma_{XC} = \gamma^*_{XC}(\beta(c_E - c_T)/(CP + P_E))$
2	Cross-compliance	$P_E = 0$	$P = CP$	$\gamma_X = \gamma^*_X(\beta(c_E - c_T)/(CP))$
3	Command and control	$P_E = 0$	$P = P_L$	$\gamma_{CC} = \gamma^*_{CC}(\beta(c_E - c_T)/(P_L))$
4	Incentive scheme	$P_E = C_E - C_T$	$P = P_E$	$\gamma_I = \gamma^*_I(\beta(c_E - c_T)/(P_E))$

MC = monitoring costs.

would select farmers in accordance to their actual C_E and would also help to achieve a socially optimum K^* (Baneth, 1994).

Based on the previous discussion and our theoretical results, it can be argued that the incentive–penalty scheme seems to be the most appropriate policy alternative to the current incentive approach. Introducing the current parameter values established under the cereal CMO and the agri-environmental programme in Spanish cereal lands, the difference between the monitoring costs of both approaches will be estimated .

Thus, given the following parameters:[3]

c_m (monitoring cost) = 3.28 European currency units (ECU) ha^{-1}
p_e (current '2078/92' payment) = c_e = 34.75 ECU ha^{-1}
r (percentage set-aside) = 10
a (percentage cultivated land) = 60.9
$1 - \alpha - \alpha r$ (percentage WF index) = 33
p_c (cereal compensatory payment) = 106.25 ECU ha^{-1}
p_r (set-aside compensatory payment) = 137.5 ecu ha^{-1}
c_T (WF cost) = 15.62 ECU ha^{-1}
w (market returns for cereal land) = 62.5 ECU ha^{-1}
with $P_E = p_e(1 - \alpha - \alpha r)$; $CP = p_c\alpha + p_r(1 - \alpha)$; $W = w\alpha$;
$C_T = c_t(1 - \alpha - \alpha r)$

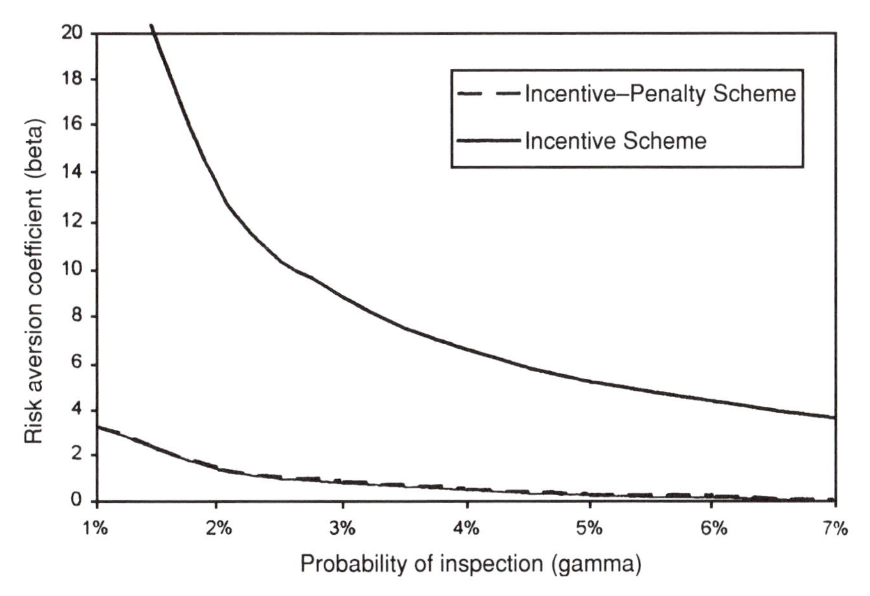

Fig. 17.1. Trade-off beta–gamma curves for incentive versus incentive–penalty schemes.

we can now plot γ_I^* and γ_{IP}^* for a plausible range of risk-aversion coefficients. Figure 17.1, then, shows the relationship between γ^* and β for both policy options. As expected, it is confirmed that the optimum probability of inspection decreases as the risk-averse coefficient increases.

A simple look at Fig. 17.1 makes the difference between the incentive–penalty and incentive schemes quite meaningful. Given any risk-averse attitude of farmers, it becomes clear that the optimum probability of inspection (γ^*) has to be set at considerably higher levels than the reference level pursuant to Council Regulation (EEC) 746/96 when an incentive scheme is implemented. It can be concluded, then, that the implementation of the current '2078/92' programme – which follows the incentive scheme – is challenged with the difficulty of setting monitoring levels well over the Spanish administrative capacity.

Figure 17.2 shows the resultant monitoring cost for different values of β and the remarkable difference between incentive–penalty and incentive schemes. Particularly, monitoring costs of incentive schemes stand seven or eight times higher than monitoring costs of incentive–penalty schemes for any given value of β.

Furthermore, our results suggest that, under the current incentive-scheme approach, farmers might enrol in the agri-environmental programme, as it satisfies the rationality constraint, but very few farmers – only those with an extremely risk-averse utility function – would be likely to comply with the established requirements, given the current level of inspection, γ, set at 5%.

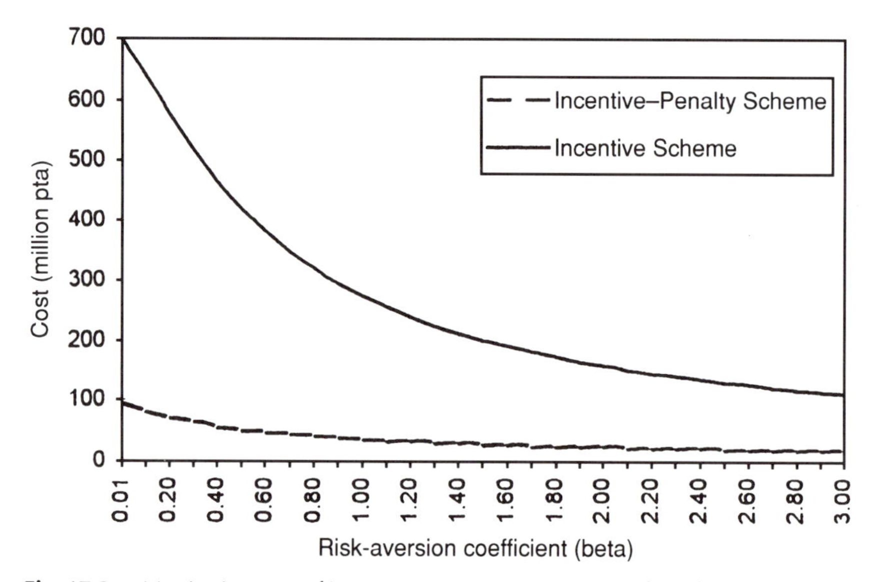

Fig. 17.2. Monitoring cost of incentive versus incentive–penalty schemes. pta, Pesetas.

SUMMARY AND CONCLUSIONS

As was stated at the beginning of this chapter, the agri-environmental Council Regulation '2078/92' has opened new opportunities to integrate environmental objectives within the CAP reform. Several research works have been carried out to assess the cost-efficiency of different policy instruments to implement agri-environmental programmes, among which those of Crabtree and Chalmers (1994) and Richard and Trommetter (1994) deserve special mention. These works are mainly focused on the problem of situations where farmers taking up the same measure might face different costs and on the optimum uptake response by farmers. Our approach differs in that we tackle the problem of hidden information, about the farmer's actions, which cannot be directly observed except with high monitoring costs, which arises in most of the agri- environmental programmes.

We have tried to assess the agri-environmental programme related to extensification measures in Spanish cereal lands and establish qualitative results about the different monitoring costs of alternative policy options. A new policy instrument has been proposed, namely an incentive–penalty scheme, which ensures for the risk-averse farmer a fixed payment – incentive – for undertaking a given set of environmental practices, but provides a much riskier setting for farmers reluctant to comply with the contract's terms than the current incentive scheme.

A primary qualitative result reached in the analysis is that monitoring costs are highly influenced by the selected policy option and are directly related to the risk implied by the payment schedule the farmer faces if he/she decides to commit fraud. Applying our theoretical model to the specific conditions of the implementation of Council Regulation (EEC) 2078/92 in Spanish traditional fallow, it becomes clear that the current approach – which follows an incentive scheme – faces remarkable difficulties, due to the fact that farmers are encouraged to enter the programme but are not sufficiently induced to comply with the established requirements. We also show that the optimum probability of inspection would have to be set at levels well over the Spanish administrative capacity. The results indicate that an incentive–penalty approach overcomes this drawback and could lower by several times the monitoring costs of the programme.

While our investigation suggests a promising way of assessing policy alternatives for integrating environmental objectives within the CAP reform, we acknowledge that our stylized analysis should be further developed to integrate issues that have already been examined by other authors. Further steps could be taken to introduce both dynamic and stochastic aspects related to the variability of farmers' costs within a more realistic theoretical framework.

NOTES

1 Proofs of solutions to the model for each policy scheme are available from the authors upon request.

2 Taking as a reference point the monetary fines established by several legislative bodies, we can assume that $P_E < P < P_{CMO}$

3 These figures were facilitated by the Spanish Ministry of Agriculture for the years 1995 and 1996, except parameter w, which was estimated by the authors.

Research Results and Policy Implications

18

Stephan Dabbert, Alex Dubgaard, Louis Slangen and Martin Whitby

INTRODUCTION

During more than a decade, the public debate on deterioration of landscape quality and the decrease in wildlife and biodiversity in Europe resulting from agricultural and other land-management practices has led to policy action. In the context of the agricultural Common Agricultural Policy (CAP) reform of 1992, environmental considerations became for the first time an integrated part of European agricultural policy, building on and extending the tentative policy measures introduced in the 1980s. This by no means implies that the debate on the environmental impact of agriculture and on how best to deal with it is over. On the contrary, future policy developments are expected that will put more emphasis on this aspect than at present.

The debate on the issues of wildlife and landscape conservation was fuelled initially by arguments taken from the natural sciences. During the past decade, however, the amount of social science research in this area has increased considerably. The earlier chapters of this book show the diversity of the work done, including contributions from sociologists, political and natural scientists. This final chapter attempts to review the main results of the research presented at the workshop and to discuss the policy implications from them.

ANALYTICAL ISSUES

The main disciplinary background of those presenting chapters has been from the social sciences, in particular economics. Mainstream (neoclassical) welfare

 The Economics of Landscape and Wildlife Conservation (eds S. Dabbert, A. Dubgaard, L. Slangen and M. Whitby)

economics is the theoretical base for most of the present analyses of nature- and landscape-conservation problems. Generally, welfare economics assumes that individuals are self-interested agents who make rational choices in accordance with their preferences. Hence, the welfare concept is that of preference satisfaction. The formation of individual preferences is not seen as a subject of economic investigations. Similarly, defining the appropriate institutional structure of the economy/society is considered as a political question beyond the scope of economic analysis. It is generally assumed that institutions like governments and administrative agencies are neutral agents seeking the 'best' for society as a whole.

These positions are challenged in the chapters by Whittaker, Osti and Holstein (Chapters 2, 3 and 4), questioning the theoretical framework as well as the value assumptions of mainstream economics. Whittaker presents the positions of ecological economics and advocates a multidisciplinary approach to environmental-policy analysis, where natural sciences are integrated with social sciences. In a chapter focusing on the intricacies of establishing national parks in the River Po delta, Osti contests the policy-formation assumptions made by neoclassical economics. Rather, institutional interests are seen as the determining factors in the policy-making process. Both Whittaker and Holstein question the value assumptions made by welfare economics. Preferences/values are moulded by their social context and choice is limited by the structure of the economy/society. Individuals may have other value perspectives than those attributed to the self-interested economic agent of neoclassical welfare economics. If so, this could be critical for the assumptions made by one of the new fast-growing branches of environmental economics – economic valuation, which seeks to measure environmental benefits in terms of money.

The debate during the workshop did not suggest how, or if, the positions of neoclassical economics and the alternative approaches can be reconciled. But the contributions by Whittaker, Osti and Holstein elucidated the problems of the value judgements involved in moving from economic conclusions to policy implications. Hence, the value judgements on which the conclusions of economic analysis are based should be clearly stated for policy-makers.

ECONOMIC VALUATION AND POLICY APPRAISAL

Environmental benefits are generally non-market and decisions about provision and preservation are made through the political and administrative system. Scarcity of resources implies that environmental objectives must be balanced against other social ends, and decision-makers have to find ways of weighing the desirability of different alternatives. Cost–benefit analysis (CBA) provides a framework for this type of policy assessment. However, application of CBA in environmental-policy analyses is often limited by the lack of monetized estimates of environmental values. As noted above, this has led to a rapid development of methods to measure environmental benefits in terms of money. The chapters by Dubgaard and Hasund (Chapters 5 and 6), respectively, deal with the valuation of benefits from landscape amenities and outdoor recreation. Other chapters make use of the results of such studies: those by Bonnieux and Le Goffe, Oskam and Slangen, and Defrancesco and Merlo (Chapters 7, 9 and 10, respectively). The conclusions from the economic-valuation studies indicate that it is possible to make consistent and fairly robust assessments of individuals' preferences (in the form of willingness to pay) for landscape benefits.

The classical economic instrument for policy appraisal is CBA. The two examples here, by Oskam and Slangen, and Bonnieux and Le Goffe (Chapters 9 and 7), demonstrate the broad applicability of the technique, in that the first concerns a major land-use change project in The Netherlands and the second a change in policy regarding the restoration of hedgerows in France. Defrancesco and Merlo (Chapter 10) take a methodologically somewhat different approach, including landscape and environmental values in an individual enterprise-accounting framework.

Economic-valuation studies as well as landscape-policy analyses generally support the scientific claims for the importance of the environment, but the number of methodological caveats attached to most empirical work in this field suggests that policy-makers may not yet be given a crisp answer to the question, 'How important is the environment?', which they may take as unassailable truth. Rather, the results from such studies should be seen as a contribution to informed debate and policy decision-making.

INSTITUTIONS

Institutional questions are highly significant for policy-makers and never more so than in environmental policy. Institutions are particularly important here because environmental problems are typically seen as a result of market failure. The usual response to such failure is some form of policy intervention and typically this has institutional implications. These may take the form of significant transactions costs (Whitby *et al.*, Chapter 8), which are likely to be particularly important if the environmental outcomes are ill-defined and difficult to measure. This is typically the case when these outcomes arise from a numerous constituency of actors, namely farmers. However, Whitby *et al.* point out that policies seeking to reduce the intensity of input use can also be expected to bring the benefits of reduced surplus production.

The quest for sensible environmental-policy instruments has provoked innovation in the design and implementation of policies, a key issue being the identification of appropriate rates of payment. Most of the new institutional arrangements consist of some form of contract between farmers and the state, where a 'price' is 'posted' defining the reward for policy compliance and farmers come forward to make contracts. An important general source of market failure arises from informational asymmetries. In such cases, we can only choose a second best policy (Weaver, Chapter 11). Asymmetric information gives rise to opportunistic behaviour through adverse selection and moral hazard (Latacz-Lohmann, Chapter 12): problems that may be offset by well- designed auction systems. The appropriate price may also be identified by direct negotiation, which is used in arriving at some forms of contract. The use of the principal–agent model is further applied by Sumpsi *et al.* (Chapter 17). These models all cast governments in the role of buyers of wildlife and landscape from farmers, who are seen as their producers. Various other contributors referred to the problems of designing appropriate institutions and policies. For example, Kazenwadel *et al.* (Chapter 13) see government setting the price and 'supplying' agri-environment schemes. Weaver (Chapter 11) makes an interesting contribution to institutional design, describing a private system of provision of public environmental goods.

Other studies focus more on modelling the implications of policy, with examples from The Netherlands (Wossink *et al.*, Chapter 14) and Germany (Piorr *et al.*, Chapter 15) and a review of the management of such research projects (Moxey and White, Chapter 16). Such spatial models have to combine a large amount of physical and financial information and produce normative estimates of optimal resource allocation. They have been developed mainly during the 1990s and their use by policy-makers as a decision aid has been slight so far.

FUTURE PROSPECTS

The future of landscape and wildlife on agricultural and forest land will depend on the types of policy introduced and their effectiveness. At the time of writing, there has been substantial interest in the implementation of Council Regulation (European Economic Community (EEC)) 2078/92 and its successor Council Regulation (EEC) 746/96, and many member states have implemented schemes. But it remains the case that most of the instruments introduced are voluntary and that it will require several years for policy benefits to appear. Meanwhile, the budgets for these policies and their general objectives are known. Only a few policies used have a more direct influence on the environment, which can be assessed directly.

Most of these policies have been introduced as simple additional instruments in the array of policies confronting farmers. Thus, many farmers will have to choose between responding to market signals that may encourage them to intensify their production systems and adopting one of an array of policies that will press them to reduce the use of certain inputs or change their management in other prescribed ways.

The need for consistent and efficient landscape-policy programmes has involved some days of intense discussion at the University of Hohenheim, followed by selection and intensive editing of half the papers presented here for this publication. They are offered as a contribution to future debates in this vital and dynamic policy field.

References

Adams, R., Dunn, S.M., Lunn, R., Mackay, R. and O'Callaghan, J.R. (1995) Assessing the performance of the NELUP hydrological models for river basin planning. *Journal of Environmental Planning and Management* 38, 53–76.

Adger, W.N. and Whitby, M.C. (1993) Natural resource accounting in the land-use sector: theory and practice. *European Review of Agricultural Economics* 20, 77–97.

AdL (Akademie der Landwirtschaftwissenschaften der DDR) (1989) *Datenspeicher 'Sozialistische Landwirtschaft'*. AdL, Böhlitz-Ehrenberg.

Agostino, M. (1993) Finanziamenti per l'agricoltura biologica nei parchi nazionali. *Bioagricoltura* 22, 9–12.

Agriculture Committee (1997) *Environmentally Sensitive Areas and Other Schemes under the Agri-Environment Regulation*, Vol. I. House of Commons, London.

Alexander, J.C. (1995) I paradossi della società civile. *Rassegna Italiana di Sociologia* 3, 319–339.

Allanson, P., Moxey, A.P. and White, B. (1993) Measuring agricultural non-point pollution for river catchment planning. *Journal of Environmental Management* 38, 219–232.

Allen, T.F.H. and Starr, T. (1982) *Hierarchy: Perspectives for Ecological Complexity*. University of Chicago Press, Chicago.

Alonso, W. (1964) *Location and Land Use: Toward a General Theory of Land Rent*. Harvard University Press, Cambridge, Massachusetts.

Alston, J.M. and Hurd, B.H. (1990) Some neglected social costs of government spending in farm programs. *American Journal of Agricultural Economics* 72, 149–156.

von Alvensleben, R. (1995) Naturschutz im Lichte der Standorttheorie. *Agrarwirtschaft* 44, 230–236.

Anderberg, T. (1994) *Den mänskliga naturen. En essä om miljö och moral*. Nordstedts, Värnamo, 302 pp.

Andrews, R.N.L. and Waites, M.J. (1978) *Environmental Values in Public Decisions: a Research Agenda*. School of Natural Resources, University of Michigan, Ann Arbor.

Anon. (1987) *Kemp Seminar: Improving Interdisciplinary Research: Integrating the Social and Natural Sciences.* Center for Resource Policy Studies, School of Natural Resources, University of Wisconsin–Madison.

Antle, J.M. and McGuckin, T. (1993) Technological innovation, agricultural productivity and environmental quality. In: Carlson, G., Zilberman, D. and Miranowski, J.A. (eds) *Agricultural and Environmental Resource Economics.* Oxford University Press, New York, pp. 175–220.

Archambault, E. and Bernard, J. (1988) *Système de Comptabilité de l'Environnement et Problèmes d'Evaluation Économique: l'Approche Française.* Laboratoire d'Economie Sociale, Paris.

Archibald, S. (1988) Incorporating externalities into agricultural productivity analysis. In: Capalbo, S. and Antle, J. (eds) *Agricultural Productivity Measurement and Explanation.* Resources for the Future, Washington, DC, pp. 366–393.

Ariansen, P. (1993) *Miljöfilosofi.* Nya Doxa, Nora, 231 pp.

Aronsson, M., Hallingbäck, T. and Mattsson, J.-E. (eds) (1995) *Swedish Red Data Book of Plants 1995.* Artdatabanken, Threatened Species Unit, Swedish University of Agricultural Sciences, Uppsala.

Arrow, K., Solow, R., Portney, P.R., Leamer, E.E., Radner, R. and Schuman, H. (1993) Report of the NOAA Panel on Contingent Valuation. Report to the General Council of the United States National Oceanic and Atmospheric Administration. *US Federal Register* 58(10), 4602–4614.

Arrow, K., Bolin, B., Costanza, C., Dasgupta, P., Folke, C., Holling, C.S., Jansson, B.O., Levin, S., Maler, K.-G., Perrings, C. and Pimentel, D. (1995) Economic growth, carrying capacity, and the environment. *Science* 268, 520–521. (Reproduced in *Ecological Economics* 15, 91–95.)

van Asperen, H.S.B.M. (1994) Evaluatie van het Natuurbeleid. In: Silvis, H.J. (ed.) *Beleid en Universiteit* 5. Werkgroep Landbouwpolitiek, Wageningen, pp. 9–25.

Bagnasco, A. and Trigilia, C. (eds) (1984) *Società e Politica nelle Aree di Piccola Impresa.* Arsenale, Venice, 315 pp.

Baldock, D. (1992) The polluter pays principle and its relevance to agricultural policy in the European countries. *Sociologia Ruralis* 32, 49–65.

Baldock, D. and Mitchell K. (1995) *Cross-Compliance within the Common Agricultural Policy.* Institute for European Environmental Policy, London.

Ballard, C.L., Shoven, J.B. and Whalley, J. (1985) General equilibrium computations of the marginal welfare costs of taxes in the United States. *American Economic Review* 75, 128–138.

Baneth, M.-H. (1994) Auctions as a means of creating a market for environmental services in the countryside. Paper presented at the Workshop on Agricultural Policy and the Countryside, Oslo, 8–9 December.

Barlow, R. (1972) *Land Resource Economics: the Economics of Real Property*, 2nd edn. Prentice Hall, London, 616 pp.

Bartolomeo, M., Malaman, R., Pavan, M. and Sammarco, G. (1995) *Il Bilancio Ambientale d'Impresa.* Il sole 24 ore-Pirola, Milan.

Basile, E. and Cecchi, C. (1995) Dal declino dell'agricoltura alla formazione dei sistemi locali rurali. In: CESAR (ed.) *Il Sistema di Agrimarketing e le Reti d'Impresa.* Università degli Studi di Perugia, Perugia, pp. 269–296.

Bateman, I. (1994) Research methods for valuing environmental benefits. In: Dubgaard, A., Bateman, I. and Merlo, M. (eds) *Economic Valuation of Benefits from Countryside Stewardship.* Wissenschaftsverlag Vauk, Kiel, pp. 47–82.

Bateman, I., Garrod, G., Diamond, E. and Langford, I. (1995) *An Economic Appraisal of the Viability of the Community Woodland Scheme.* Working Paper 15, CRE, University of Newcastle upon Tyne.

Bateman, I.J. and Turner, R.K. (1993) Valuation of the environment, methods and techniques: the contingent valuation method. In: Turner, R.K. (ed.) *Sustainable Environment Economics and Management: Principles and Practice.* Belhaven Press, London, pp. 120–191.

Batie, S.S. (1989) Sustainable development: Challenges to the profession of agricultural economics. *American Journal of Agricultural Economics* 71, 1083–1101.

Batie, S.S. (1992) Sustainable development: concepts and strategy. In: Peters, G.H. and Stanton, B.F. (eds) *Proceedings of the Twenty-First Conference of Agricultural Economist.* Dartmouth Publishing, Aldershot, pp. 391–402.

Baumol, W.J. and Oates, W.C. (1975) *The Theory of Environmental Policy.* Prentice Hall, Englewood Cliffs, New Jersey.

Bawden, R.J. and Ison, R.L. (1992) The purpose of field-crop ecosystems: social and economic aspects. In: Pearson, C.J. (ed.) *Field Crop Ecosystems.* Elsevier, London, pp. 11–35.

Beasley, S.D., Workman, W.G. and William, N.A. (1986) *Non-market Valuation of Open Space and Amenities Associated with Retention of Lands in Agricultural Use.* Bulletin 71, Agricultural and Forestry Experiment Station, University of Alaska – Fairbanks.

Beaufoy, G. (1994) Impact of the CAP reform on land use and rural amenities. In: Dubgaard, A., Bateman, I. and Merlo, M. (eds) *Economic Valuation of Benefits from Countryside Stewardship.* Wissenschaftsverlag Vauk, Kiel.

Benvenuti, B. (1975) General system theory and entrepreneurial autonomy in farming. *Sociologia Ruralis* 1–2, 46–61.

van Bergeijk, P.A.G., van Sinderen, J. and Waasdorp, P.M. (1993) De dynamiek van belastingverlaging. *Openbare Uitgaven* 23, 23–29.

van Bergen, J.A.M. and Biewenga, E.E. (1992) *Landbouw en Broeikaseffect.* CLM rapport 84, Utrecht.

Bergland, O., Fried, B.M. and Adams, R.M. (1996) *Temporal Stability of Values from the Contingent Valuation Method.* Report D-03/1996, Department of Economics and Social Sciences, Agricultural University of Norway, Ås.

Bernes, C. (ed.) (1994) *Biological Diversity in Sweden: a Country Study.* Monitor 14, Swedish Environmental Protection Agency, Solna.

Berrens, R.B. and Polansky, S. (1995) The Paretian liberal paradox and ecological economics. *Ecological Economics* 14, 45–46.

Billaud, J.-P. (1986) L'etat nécessaire? Aménagement et corporatisme dans le marais poitevin. *Etudes Rurales* 101–102, 73–111.

BirdLife International (1996) *Nature Conservation Benefits of Plans under the Agri-Environment Regulation (EEC 2078/92).* Project co-funded by DG XI of the European Commission and carried out by BirdLife International (contract B4–3040/94/239/AO/B2) BirdLife International, Cambridge.

Blamey, R., Common, M. and Quiggin, J. (1995) Respondents to contingent valuation surveys: consumers or citizens. *Australian Journal of Agricultural Economics* 39, 263–288.

Boadway, R.W. and Bruce, N. (1984) *Welfare Economics*. Basil Blackwell, Oxford, 344 pp.

Bonnieux, F. and Rainelli, P. (1995) Contingent valuation and the design of agri-environmental measures. In: Hofreither, M.F. and Vogel, S. (eds) *The Role of Agricultural Externalities in High Income Countries*. Wissenschaftsverlag Vauk, Kiel, pp. 91–108.

Bonnieux, F. and Weaver, R. (1996) Environmentally sensitive area schemes: public economics and evidence. In: Whitby, M. (ed.) *The European Environment and CAP Reform, Policies and Prospects for Conservation*. CAB International, Wallingford, pp. 209–226.

Bonnieux, F., Rainelli, P. and Vermersch, D. (1995) The provision of environmental goods by agriculture. Paper presented to the Sixth Annual Conference of the European Association of Environmental and Resource Economists, Umeå, Sweden, 18–20 June.

Boots, M., Oude Lansink, A. and Peerlings, J. (1997) Efficiency loss due to distortions in Dutch milk quota trade. *European Review of Agricultural Economics* 24, 31–46.

Bork, H.-R., Dalchow, C., Kächele, H., Piorr, H.-P.and Wenkel, K.O. (1995) *Agrarlandschaftswandel in Nordost-Deutschland unter veränderten Rahmenbedingungen: ôkologische und ôkonomische Konsequenzen*. Ernst & Sohn Verlag, Berlin, 418 pp.

Bossi, C. (1992) L'audit ambientale aiuta le imprese. *L'Impresa Ambiente* 1, 38–41.

Boumans, R.M.J. and Sklar, F.H. (1990) A polygon-based spatial (PBS) model for simulating landscape change. *Landscape Ecology* 4, 83–97.

Bouwhuis, T. (1995) *Weidevogelbescherming in Waterland 1995*. Samenwerkingsverband Waterland, Purmerend.

Bromley, D.W. (1991) *Environment and Economy: Property Rights and Public Policy*. Basil Blackwell, Oxford.

Brooks, F.P. (1975) *The Mythical Man-month: Essays on Software Engineering*. Addison-Wesley, Reading, Massachusetts.

Brouwer, F.M. and van Berkum, S. (1996) *CAP and the Environment in the European Union: Analysis of the Effects of CAP on the Environment and an Assessment of Existing Environmental Conditions in Policy*. Wageningen Pers, Wageningen, 171 pp.

Brouwer, R. and Slangen, L.H.G. (1995) *The Measurement of the Non-Marketable Benefits of Agricultural Wildlife Management: The Case of Dutch Peat Meadow Land*. Department of Agricultural Economics and Policy, Agricultural University, Wageningen, 51 pp.

Brown, T.C. (1984) The concept of value in resource allocation. *Land Economics* 60, 231–246.

Browning, E.K. (1987) On the marginal welfare costs of taxation. *American Economic Review* 77, 11–23.

Brunori, G. (1994) Spazio rurale e processi globali: alcune considerazioni teoriche. In: Panattoni, A. (ed.) *La Sfida della Moderna Ruralità*. Cnr-Raisa, Pisa, pp. 1–25.

Brusco, S. and Natali, A. (1992) *Una Analisi per Aree dell'Economia Ferrarese*. Eco & Eco, Bologna, 29 pp.

Bundesregierung (1992) *Agrarbericht*. Bundesregierung, Bonn.

Buys, J.C. (1995) *Towards a Yardstick for Biodiversity on Farms* (in Dutch with English summary). CLM 169, Centrum voor Landbouw en Milieu, Utrecht.

Carlson, G.A., Zilberman, D. and Miranowski, J.A. (eds) (1993) *Agricultural and Environmental Resource Economics*. Oxford University Press, Oxford.

Carson, R.T., Mitchell, R., Hanemann, M.W., Kopp, R.J., Presser, S. and Rudd, P.A. (1992) *A Contingent Valuation Study of Lost Passive Use Values Resulting from the Exxon Valdez Oil Spill*. Attorney General of the State of Alaska, Anchorage.

Castiglione, C. (1994) Agricoltura e aree protette. In: Unione delle Province Italiane (ed.) *Parchi e Province. Istituzione, Programmazione e Gestione*. Supplement of *Le Autonomie* 9, 31–43.

Ciriacy-Wantrup, S.V. (1952) *Resource Conservation: Economics and Policies*. Division of Agricultural Sciences, University of California. Berkeley, California.

Colman, D. (1991) Land purchase as a means of providing external benefits from agriculture. In: Hanley, N. (ed.) *Farming and the Countryside: an Economic Analysis of External Costs and Benefits*. CAB International, Wallingford, pp. 215–229.

Colman, D.R., Crabtree, J.R., Froud, J. and O'Carroll, L. (1992) *Comparative Effectiveness of Conservation Mechanisms*. Department of Agricultural Economics, University of Manchester, Manchester, UK.

Colson, F. and Stenger-Letheux, A. (1996) Evaluation contingente et paysages agricoles. *Cahiers d'economie et Sociologie Rurales* 30–40, 151–179.

Commissione delle Comunità Europee (1988) Il futuro del mondo rurale. Supplement of *Rivista di Politica Agraria Italiana* 4, 20.

Commission of the European Communities (1995) *Study on Alternative Strategies for the Development of Relations in the Field of Agriculture between the EU and the Associated Countries with a view to future Accession of these Countries (Agricultural Strategy Paper)* CSE (95) 607. Commission of the European Community, Brussels, Luxemburg.

Commission of the European Communities (1996) *Progress Report from the Commission on the Implementation of the European Community Programme of Policy and Action in Relation to the Environment and Sustainable Development 'Towards Sustainablility'*. COM (95) 624 final, Office for Official Publications of the European Communities, Luxemburg.

Common, M. (1995) *Sustainability and Policy – Limits to Economics*. Cambridge University Press, Cambridge.

Communautés Européennes (1985) Décision de la Commission, du 7 juin 1985, portant établissement d'une typologie communautaire des exploitations agricole (85/377/CEE). *Journal Officiel des Communautés Européennes* L220, 1–32.

Conway, G.R. (1987) The properties of agroecosystems. *Agricultural Systems* 24, 95–118.

Cooper, J.C. (1993) Optimal bid selection for dichotomous choice contingent surveys. *Journal of Environmental Economics and Management* 13, 255–268.

Cooper, J.C. and Loomis, J.B. (1993) Sensitivity of willingness to pay estimates to bid design in dichotomous choice contingent valuation models. *Land Economics* 60, 203–239.

Costanza, R. and Patten, B.C. (1995) Defining and predicting sustainability. *Ecological Economics* 15, 193–196.

Costanza, R., Daly, H.E. and Bartholomew, J.A. (1991) Goals, agenda and policy recommendations for ecological economics. In: Constanza, R. (ed.) *Ecological Economics:*

the Science and Management of Sustainability. Columbia University Press, New York, pp. 1–20.

Costanza, R., Segura, O. and Martinez-Alier, J. (eds) (1996) *Getting Down to Earth: Practical Applications of Ecological Economics*. Island Press, Washington, DC.

Council Regulation (EEC) 797/85 of 30 March 1985 on improving the efficiency of agricultural structures. *Official Journal of the European Communities* L93, 1–18.

Council Regulation (EEC) 1760/87 of 15 June 1987 amending Regulations (EEC) No. 797/85, (EEC) No. 270/79, (EEC) No. 1360/78 and (EEC) No. 355/77 as regards agricultural structures, the adjustment of agriculture to the new market situation and the preservation of the countryside. *Official Journal of the European Communities* L167, 1–8.

Council Regulation (EEC) 2092/91 of 24 June 1991 on organic production of agricultural products and indications referring thereto on agricultural products and foodstuffs. *Official Journal of the European Communities* L198, 1–15.

Council Regulation (EEC) 2078/92 of 30 June 1992 on agricultural production methods compatible with the requirements of the protection of the environment and the maintenance of the countryside. *Official Journal of the European Communities* L215, 85–90.

Council Regulation (EEC) 2085/93 of 20 July 1993 amending Regulation (EEC) No. 4256/88 laying down provisions for implementating of Regulation (EEC) No 2052/88 as regards the European Agricultural Guidance and Guarantee Fund (EAGGF) Guidance Section. *Official Journal of the European Communities* L193, 44–47.

Council Regulation (EEC) 746/96 of 24 April 1996 laying down detailed rules for the application of Council Regulation (EEC) No 2078/92 on agricultural production methods compatible with the environment and the maintenance of the countryside. *Official Journal of the European Communities* L102, 19–27.

Cox, G., Lowe, P. and Winter, M. (1986) From state direction to self regulation: the historical development of corporatism in British agriculture. *Policy and Politics* 4, 475–490.

Crabtree, J.R. and Chalmers, N.A. (1994) Economic evaluation of policy instruments for conservation. *Land Use Policy* 11, 94–106.

Crespi, F. (1989) La fine della secolarizzazione: dalla sociologia del progresso alla sociologia dell'esistenza. In: Mongardini, C. and Maniscalco, M.L. (eds) *La Sociologia del Postmoderno*. Bulzoni, Rome, pp. 153–163.

Curtis, L. (1983) Reflections on management agreements for conservation of Exmoor moorland. *Journal of Agricultural Economics* 34, 397–406.

Dannowski, R. (1995) Landschaftsindikator Wasserbilanz. In: Bork, H.-R., Dalchow, C., Kächele, H., Piorr, H.-P. and Wenkel, K.O. (eds) *Agrarlandschaftswandel in Nordost-Deutschland unter veränderten Rahmenbedingungen: Ökologische und Ökonomische Konsequenzen*. Ernst & Sohn Verlag, Berlin, pp. 149–165.

Dasgupta, P., Kriström, B. and Maler, K.G. (1995) Current issues in resource accounting. In: Johansson, P.O., Kriström, B. and Maler, K.G. (eds) *Current Issues in Environmental Accounting*. Manchester University Press, Manchester, pp. 117–152.

De Backer, D. (1992) *Le Management Vert*. Dunot, Paris.

Defrancesco, E. and Merlo, M. (1996) L'esposizione dei beni e servizi ambientali nel bilancio dell'azienda forestale. *Genio Rurale* 7/8, 50–62.

Dente, B. (1985) Centre–local relations in Italy: the impact of the legal and political structures. In: Meny, Y. and Wright V. (eds) *Centre–Periphery Relations in Western Europe.* Allen & Unwin, London, pp. 125–148.

Dente, B. and Ranci, P. (1992) *L'Industria e l'Ambiente.* Il Mulino, Bologna.

Department of the Environment (1993) *Environmental Appraisal of Development Plans: a Good Practice Guide.* HMSO, London.

De Schutter, A.J.M. (1993) Ontwikkelingen in de grondverwerving. *Landinrichting* 33, 25–27.

Diamond, P.A. and Hausman, J.A. (1994) Contingent valuation: is some number better than no number? *Journal of Economic Perspectives* 8, 45–64.

Di Iacovo, F. (1994) Le politiche agricole e lo sviluppo locale. In: Panattoni, A. (ed.) *La Sfida della Moderna Ruralità.* Cnr-Raisa, Pisa, pp. 127–211.

Dillman, B.L. and Bergstrom, J.C. (1991) Measuring environmental amenity benefits of agricultural land. In: Hanley, N. (ed.) *Farming and the Countryside.* CAB International, Wallingford, pp. 250–271.

van Dorp, D. (1996) Seed dispersal in agricultural habititats and the restoration of species-rich meadows. PhD thesis, Wageningen Agricultural University, Wageningen.

Dosi, G. (1988) Sources, procedures, and microeconomic effects of innovation. *Journal of Economic Literature* 26, 1120–1171.

Dow, S.C. (1997) Mainstream economic methodology. *Cambridge Journal of Economics* 21, 73–93.

Drake, L. (1987) *The Value of Preserving the Agricultural Landscape – Result from Surveys.* Report 289, Department of Economics and Statistics, Swedish University of Agricultural Sciences, Uppsala, 67 pp.

Drake, L. (1992) The non-market value of the Swedish agricultural landscape. *European Review of Agricultural Economics* 19, 351–364.

Drummond, I. and Symes, D. (1996) Rethinking sustainable fisheries: the realistic paradigm. *Sociologia Ruralis* 36, 152–162.

Dubgaard, A. (1994) Valuing recreation benefits from the Mols Bjerge area, Denmark. In: Dubgaard, A., Bateman, I. and Merlo, M. (eds) *Economic Valuation of Benefits from Countryside Stewardship.* Wissenschaftsverlag Vauk, Kiel, pp. 145–163.

Dubgaard, A. (1996a) *Economic Valuation of Recreation in Mols Bjerge.* SØM publikation 11, AKF Forlaget (Institute of Local Government Studies), Copenhagen.

Dubgaard, A. (1996b) Policies for landscape and nature conservation in Denmark. In: Umstätter, J. and Dabbert, S. (eds) *Policies for Landscape and Nature Conservation in Europe.* An Inventory to Accompany the Workshop on 'Landscape and Nature Conservation' held on 26–29 September 1996 at the University of Hohenheim, Germany. Department of Farm Management, Production Theory (and Resource Economics Unit), University of Hohenheim, Stuttgart.

Dubgaard, A. (1997) *Contingent Valuation Survey of Willingness to Pay for Access to Danish Forest.* SØM publikation 11, AKF Forlaget (Institute of Local Government Studies), Copenhagen.

Dubgaard, A., Bateman, I. and Merlo, M. (eds) (1994) *Economic Valuation of Benefits from Countryside Stewardship.* Wissenschaftsverlag Vauk, Kiel.

Duelli, P. (1992) Mosaikkonzept und Inseltheorie in der Kulturlandschaft. *Verhandungender Gesellschaft Ökologie* 21, 379–384.

Duffield, J.W. and Patterson W.E. (1991) Inference and optimal design for a welfare measure in dichotomous choice contingent valuation. *Land Economics* 67, 225–239.

Dutton, J.M. and Starbuck, W.H. (1971) *Computer Simulation of Human Behaviour.* John Wiley & Sons, New York.

Dye, R. (1986) Optimal monitoring policies in agencies. *Rand Journal of Economics* 17, 339–356.

Edgerton, D.L., Assarsson, B., Hummelmose, A., Laurila, I.P., Rickertsen, K. and Vale, P.H. (1996) *The Econometrics of Demand Systems.* Kluwer, Dordrecht.

Enneking, U. and Rauschmayer, F. (1996) Contingent valuation method and environmental impact assessment, ethical foundations and practical challenge. Paper presented at Colloque Ecologie, Société, Environnement, Saint-Quentin-en-Yvelines, France, 23–25 may.

European Commission (1996) *Umwelt und Regionen: für eine nachhaltige Entwicklung.* Amt für amtliche Veröffentlichungen der Europäischen Gemeinschaften, Luxemburg.

Fanfani, R. (1993) Crisi di consenso e rappresentanza nell'agricoltura italiana. *La Questione Agraria* 50, 7–13.

Farago, P. (1985) Regulating milk markets: corporatist arrangements in the Swiss dairy industry. In: Streeck, W. and Schmitter P. (eds) *Private Interest Government: Beyond Market and State.* Sage, London, pp. 178–181.

FBS (1996) *Farm Business Survey: Results for the North of England.* Department of Agricultural Economics and Food Marketing, University of Newcastle upon Tyne, Newcastle upon Tyne.

FEEM (Fondazione ENI Enrico Mattei) (1995) Company environmental reports: guidelines for preparation. Supplement to *FEEM Newsletter* 1, 1–15.

Ferro, O., Merlo, M. and Povellato, A. (1995) Valuation and remuneration of countryside stewardship performed by agriculture and forestry. In: Peters, G.H. and Hedley, D.D. (eds) *Agriculture Competitiveness: Market Forces and Policy Choice.* Dartmouth, Aldershot, pp. 415–435.

Field, B. (1994) *Environmental Economics – an Introduction.* McGraw-Hill, New York, 482 pp.

Fischler, F. (1996) Integration of environmental protection requirements into community agricultural policy and budgetary implications for 1997. Questions to Mr Fischler for the joint meeting of the Committee on Budgets and the Committee on the Environment, Public Health and Consumer Protection on 16 July 1996 in Strasburg. Unpublished.

Fisher, A.C. and Krutila, J.V. (1985) Economics of natural preservation. In: Kneese, A.V. and Sweeney, J.L. (eds) *Handbook of Natural Resource and Energy Economics*, Vol. 1. North-Holland, Amsterdam, pp. 165–189.

Fankhauser, S. (1994) *Valuing Climate Change: the Economics of the Greenhouse.* Earthscan, London.

Fraser, I. (1992) An economic analysis of compensation levels and profit foregone: implications for agri-environmental policy. PhD thesis, University of Manchester.

Fraser, I. (1995) An analysis of management agreement bargaining under asymmetric information. *Journal of Agricultural Economics* 46, 20–32.

Freeman, A.M. (1993) *The Measurement of Environmental and Resource Values, Theory and Methods.* Resources for the Future, Washington, DC.

Friedheim, T. (1992) *On the Social Benefits of Auctions.* Diskussionsschrift Nr. 26 der Forschungsstelle für Internationale Agrarentwicklung e.V., Heidelberg, 45 pp.

Frows, J. and Tatenhove, J. (1993) Agriculture, environment and the state: the development of agro-environmental policy-making in The Netherlands. *Sociologia Ruralis* 2, 220–239.

Frykblom, P. (1997) Order effects and randomized orders. In: Kriström, B. and Starrett, D. (eds) *Current Issues in Environmental Economics.* Cambridge University Press, Cambridge.

Georgescu-Roegen, N. (1979) Energy analysis and economic valuation. *Southern Economic Journal* 45, 1023–1058.

Giddens, A. (1984) *The Constitution of Society: Outline of the Theory of Structuration.* Policy Press, Cambridge.

Glemnitz, M. and Mirschel, W. (1995) Änderung im Charakter der Landnutzung. In: Bork, H.-R., Dalchow, C., Kächele, H., Piorr, H.-P. and Wenkel, K.O. (eds) *Agrarlandschaftswandel in Nordost-Deutschland unter veränderten Rahmenbedingungen: ôkologische und ôkonomische Konsequenzen.* Ernst & Sohn Verlag, Berlin, pp. 133–140.

Gratton, C. and Taylor, P. (1990) *Sport and Recreation: An Economic Analysis.* E. and F. Spon, London, 261 pp.

Gray, J. (1996) Cultivating farm life at the borders: Scottish hill sheep farms and the European Community. *Sociologia Ruralis* 36, 27–50.

Gray, R.H. (1993) *Accounting for the Environment.* Chapman and Hall, London.

Green, D., Jacowitz, K.E., Kahneman, D. and McFadden, D. (1995) Referendum contingent Valuation, Anchoring, and Willingness to Pay for Public Goods. Draft, 22 March.

Greene, W.H. (1993) *Econometric Analysis,* 2nd edn. Macmillan Publishing, New York.

Gren, I.-M., Folke, C., Turner, K. and Bateman, I. (1994) Primary and secondary values of wetland ecosystems. *Environmental and Resource Economics* 4, 55–74.

Greyson, J. (1995) *The Natural Step: a Collection of Articles.* The Natural Step in Britain, Bristol.

Grontmij (1987) *Kosten van Grondgebruik voor Natuur, Bos en Recreatie.* Grontmij, De Bilt, 26 pp.

Gueslin, A. and Hervieu, B. (1992) Un syndicalism agricole européen est-il possible? In: Hervieu, B. and Lagrave R.-M. (eds) *Les Syndicats Agricoles en Europe.* L'Harmattan, Paris, pp. 303–313.

Hampicke, U. (1991) *Natureschutz-ökonomie.* Verlag Eugen Ulmer, Stuttgart, 335 pp.

Hanemann, W.M. (1984) Welfare evaluations in contingent valuation experiments with discrete responses. *American Journal of Agricultural Economics* 66, 332–341.

Hanemann, W.M. (1994a) Valuing the environment through contingent valuation. *Journal of Economic Perspectives* 8(4), 19–43.

Hanemann, W.M. (1994b) Willingness-to-pay and willingness-to-accept: how much can they differ? *American Economic Review* 81, 635–647.

Hanley, N. (1995) The role of environmental valuation in cost–benefit analysis. In: Willis, K.G. and Corkindale, J.T. (eds) *Environmental Valuation: New Perspectives.* CAB International, Wallingford, pp. 39–54.

Hanley, N., Shogren, J.F. and White, B. (1997) *Environmental Economics in Theory and Practice.* MacMillan Press, Houndsmills, 494 pp.

Harrison, G. (1993) *General Reactions to the NOAA Report.* CVM Network.

Hasund, K.P. (1991) Jordbruket i miljöräkenskaperna- en idédiskussion. In: SOU (ed.) *Räkna med Miljön! Förslag till Natur- och Miljöräkenskaper.* Bilaga 3, Allmänna Förlaget, Stockholm, 53 pp.

Hasund, K.P. (1997) *Documentation of Three Contingent Valuation Surveys on Preservation of Landscape Elements of Agricultural Land.* Smaskriftsserien 107, Department of Economics, Swedish University of Agricultural Sciences, Uppsala

Heilbroner, R. and Milberg, W. (1995) *The Crisis of Vision in Modern Economic Thought.* Cambridge University Press, Cambridge.

Helming, J., Oskam, A. and Thijssen, G. (1993) A micro-economic analysis of dairy farming in The Netherlands. *European Review of Agricultural Economics* 20, 343–363.

Helming, J., Thijssen, G. and Oskam, A. (1995) *Decomposing Changes in Producer Behavior due to Quotas: a Micro-Economic Analysis.* University of Wageningen Press, Wageningen.

Hodge, I. (1988) Property institutions and environmental improvement. *Journal of Agricultural Economics* 39, 369–375.

Hodge, I. (1991) The provision of public goods in the countryside: how should it be arranged? In: Hanley, N. (ed.) *Farming and the Countryside: an Economic Analysis of External Costs and Benefits.* CAB International, Wallingford, pp. 179–195.

Hodgson, G.M. (1988) *Economics and Institutions.* Polity Press, Cambridge.

Holling, C.S. (1986) Resilience of ecosystems, local surprise and global change. In: Clark, W.C. and Munn, R.E. (eds) *Sustainable Development of the Biosphere.* Cambridge University Press, Cambridge, pp. 292–317.

Holling, C.S., Schindler, D.W., Walker, B.W. and Roughgarden, J. (1995) Biodiversity in the functioning of ecosystems. In: Perrings, C., Maler, K.-G., Folke, C. and Jansson, B.-O. (eds) *Biodiversity Loss: Economic and Ecological Issues.* Cambridge University Press, Cambridge, pp. 44–63.

Holt, C.A. (1980) Competitive bidding for contracts under alternative auction procedures. *Journal of Political Economy* 88, 433–445.

Hookway, R. (1967) The management of Britain's rural land. In: *Proceedings of the Town and Country Summer School.* Town Planning Institute, Belfast.

Hueting, R. (1989) Correcting national income for environmental losses: toward a practical solution. In: Ahmad, Y.J., El Serafy, S. and Lutz, E. (eds) *Environmental Accounting for Sustainable Development.* World Bank, Washington, DC, pp. 194–213.

INSEE, Ministère de l'Environnement (1986) *Les Comptes du Patrimoine Naturel.* No. 535–536 des Collections de l'INSEE, serié D, No. 137–138.

Jacowitz, K.E. and Kahneman, D. (1995) Measures of anchoring in estimation tasks. *Personality and Social Psychology Bulletin* 21 (November), 1161–1166.

Jakeman, A.J., Beck, M.B. and McAleer, M.J. (eds) (1995) *Modelling Change in Environmental Systems.* John Wiley & Sons, Chichester.

Jakobsson, K.M. and Dragun, A.K. (1996) *Contingent Valuation and Endangered Species. Methodological Issues and Application.* Edward Elgar, Cheltenham, 243 pp.

Jansen, M. (1994) *De Kosten van de Omzetting van Landbouwgrond in Natuurgebied.* Scriptie Vakgroep Algemene Agrarische Economie, Wageningen.

Jensen, F.S. and Koch, N.E. (1997) *Outdoor Recreation in (Danish) Forests 1976/77–1993/94.* Danish Forest and Landscape Research Institute, Hørsholm, Denmark.

Jensen, K.M. (1976) *Abandoned and Afforested Areas in Jutland.* Royal Danish Geographic Society, Copenhagen.

Jöbstl, H.A. (1995) *Contribution to Managerial Economics in Forestry.* Österreichischer Agrarverlag, Vienna, 88 pp.

Jones, M. (1993) Landscape as a resource and the problem of landscape values. In: Rusten, C. and Wøien, H. (eds) *The Politics of Environmental Conservation: Proceedings from a Workshop in Trondheim March 26, 1993.* Report No. 6/93, Centre for Environment and Development (SMU), University of Trondheim, pp. 19–33.

Jordbruksverket (1994) *Naturbetesmarker.* Biologisk mångfald och variation i odlingslandskapet, Stockholm, 25 pp.

Jordbruksverket (1995) *Ängar.* Biologisk Mångfald och Variation i Odlingslandskapet, Stockholm, 15 pp.

Jurgens, C.R. (1992) Tools for the spatial analysis of land and for the planning of infrastructures in multiple-land use situations. PhD thesis, Wageningen Agricultural University, Wageningen.

Jurgens, C.R. (1993) Strategic planning for sustainable rural development. *Landscape and Urban Planning* 27, 253–258.

Jurgens, C.R. (1996) *Exercises with ECONET 4.* Department of Physical Planning and Rural Development, Wageningen Agricultural University, Wageningen (in press).

Kächele, H. (1993) Modellannahmen bei der Berechnung der ökonomischen Standortqualität. In: Werner, A. and Dabbert, S. (eds) *Bewertung von Standortpotentialen im Ländlichen Raum des Landes Brandenburg.* ZALF-Berichte, Müncheberg, pp. 47–52.

Kahneman, D. and Knetsch, J.L. (1992) Valuing public goods: the purchase of moral satisfaction. *Journal of Environmental Economics and Management* 22, 57–70.

Kamminga, M.R., Hetsen, H., Slangen, L.H.G., Bisschof, N.T. and van Hoorn, A.S. (1993) *Toekomstverkenning Ruraal Grondgebruik.* NRLO-rapport 93/20, The Hague.

Kealy, M.J., Montgomery, M. and Dovidio, J.F. (1990) Reliability and predictive validity of contingent values: does the nature of the good matter? *Journal of Environmental Economics and Management* 19, 244–263.

Kemp, M.A. and Maxwell, C. (1993) Exploring a budget context for contingent valuation estimates. In: Hausman, J.A. (ed.) *Contingent Valuation. A Critical Assessment.* North-Holland, Amsterdam, pp. 217–265.

Kennan, I. and Wilson, R. (1993) Bargaining with private information. *Journal of Economic Literature* 31, 45–104.

Komen, R., Gerking, S. and Folmer, H. (1996) Income and environmental protection: empirical evidence from OECD countries. Mimeo. Wageningen.

Kristrom, B. (1995) *Practical Problems in Contingent Valuation.* Rapport 111, Sveriges Lantbruksuniversitet, Institutionen för Skogsekonomi, Umeå.

Kristrom, B. and Riera, P. (1996) Is the income elasticity of environmental improvements less than one? *Environmental and Resource Economics* 7, 45–55.

Krutilla, J.V. (1967) Conservation reconsidered. *American Economic Review* 57, 777–786.

KTBL (Kuratorium für Technik und Bauwesen in der Landwirtschaft) (1993) *Datensammlung für die Betriebsplanung in der Landwirtschaft.* KTBL, Darmstadt-Kranichstein.

Lampkin, N. (1996) Impact of EC Regulation 2078/92 on the development of organic farming in the European Union. Paper presented at CEPFAR/IFOAM Seminar on Organic Agriculture, Vignola, Italy, 6–8 June.

Lange, P. and Regini, M. (1987) Gli interessi e le istituzioni: forme di regolazione sociale e politiche pubbliche. In: Lange, P. and Regini, M. (eds) *Stato e Regolazione Sociale.* Il Mulino, Bologna, pp. 9–52.

Larini, S. (1995) I conti delle aziende vanno in verde. *Largo Consumo* 6, 148–157.

Latacz-Lohmann, U. and Van der Hamsvoort, C.P.C.M. (1996) Targeting green auctions for environmental contracting in the countryside. Paper presented at 7th Annual Conference of the European Association of Environmental and Resource Economists, Lisbon, Portugal, 26–29 June.

Latacz-Lohmann, U. and Van der Hamsvoort, C.P.C.M. (1997) Auctioning conservation contracts: a theoretical analysis and an application. *American Journal of Agricultural Economics* (in press).

Lawson, T. (1995) Realist perspective on 'economic theory'. *Journal of Economic Issues* 29, 1–32.

Le Goffe, P. (1996) Hedonic pricing of agriculture and forestry externalities. Paper presented at 7th Annual Conference of the European Association of Environmental and Resource Economists, Lisbon. Portugal, 26–29 June.

Le Grand, J. and Bartlett, W. (1993) *Quasi-Markets and Social Policy.* MacMillan, Houndsmill, 241 pp.

Lehmbruch, G. (1984) Concertazione e struttura dei 'networks' corporativi. In Lehmbruch, G. and Schmitter, P. (eds) *La Politica degli Interessi nei Paesi Industrializzati.* Il Mulino, Bologna, pp. 111–143.

LEI/CBS (Landbouw-economisch Instituut en Centraal Bureau voor de Statistiek) (1996) *Landbouwcijfers 1996.* LEI, The Hague, 273 pp.

Leipert, C. and Simonis, U.E. (1989) Environmental protection expenditure: the German example. *Rivista Internazionale di Scienze Economiche e Commerciali* 36, 255–270.

Leonard, P. (1982) Management agreements: a tool for conservation. *Journal of Agricultural Economics* 33, 351–360.

Lieberoth, I., Adler, G. and Schmidt, I. (1977) Die Nutzung der Gemeindedatei des Datenspeichers Boden in der Landwirtschaft. *Archiv für Acker und Pflanzenbau in der Bodenkunde* 21, 687–697.

Linddal, M. (1995) Forestry: environment cum economics. PhD thesis, Royal Veterinary and Agricultural University, Denmark, 186 pp.

Lindenberg, S. (1990) Homo socio-economicus: the emergence of general model of man in the social sciences. *Journal of Institutional and Theoretical Economics* 146.

Lloyd, P.E. and Dicken, P. (1978) *Location in Space: a Theoretical Approach to Economic Geography*, 2nd edn. Harper and Row, London, 474 pp.

Lockwood, M. (1997) Integrated value theory for natural areas. *Ecological Economics* 20, 83–93.

Loman, R. (1996) *Realisatie van de Ecologische Hoofdstructuur: Analyse van Beleid, Knelpunten en Grondprijzen.* Scriptie Vakgroep Algemene Agrarische Economie, Wageningen.

Long, A. and van der Ploeg, J.D. (1994) Endogenous development: practices and perspectives. In: van der Ploeg, J.D. and Long, A. (eds) *Born from Within: Practice and Perspectives of Endogenous Rural Development.* Van Gorcum, Assen.

Longworth, J.W. (1992) Presidential address: human capital formation for sustainable agricultural development. In: Peters, G.H. and Stanton, B.F. (eds) *Proceedings of the Twenty-First Conference of Agricultural Economists.* Dartmouth Publishing Company, Aldershot, pp. 18–26.

Loomis, J.B. (1989) Test–retest reliability of the contingent valuation method: a comparison of general population and visitor responses. *American Journal of Agricultural Economics* 71, 76–84.

Loomis, J.B. (1990) Comparative reliability of the dichotomous choice and open-ended contingent valuation techniques. *Journal of Environmental Economics and Management* 18, 78–85.

Lowi, T.J. (1972) Four systems of policy, politics and choice. *Public Administration Review* 32, 298–310.

Lutz, E. (1993) Toward improved accounting for the environment: an overview. In: Lutz, E. (ed.) *Toward Improved Accounting for the Environment: An UNSTAT–World Bank Symposium.* The World Bank, Washington, DC, pp. 1–14.

McAfee, R.P. and McMillan, J. (1987) Auctions and bidding. *Journal of Economic Literature* 25, 699–738.

McClean, C.J., Watson, P.M., Wadsworth, R.A., Blaiklock, J. and O'Callaghan, J.R. (1995) Land use planning: a decision support system. *Journal of Environmental Planning and Management* 38, 77–92.

McFadden, D. (1994) Contingent valuation and social choice. *American Journal of Agricultural Economics* 76, 689–708.

MAFF (1991) *Ministerial Information in MAFF (MINIM) 1990.* Ministry of Agriculture, Fisheries and Food, London.

MAFF (1992) *Ministerial Information in MAFF (MINIM) 1991/1992.* Ministry of Agriculture, Fisheries and Food, London.

MAFF (1996) *Evidence to the Select Committee on Agriculture.* 16 July, House of Commons, London.

Magleby, D.B. (1984) *Direct Legislation: Voting on Ballot Propositions in the United States.* Johns Hopkins Press, Baltimore, Maryland.

Malinvaud, E. (1972) *Lectures on Microeconomic Theory.* North-Holland, Amsterdam.

Marangoni, A. (1994) *La Gestione Ambientale.* SPACE (Security and Protection against Crime and Emergencies), Milan.

Marinelli, A., Casini, L. and Romana, D. (1990) User-benefits and the economic regional impact of outdoor recreation in a natural park of Northern Tuscany. In: Whitby, M.C. and Dawson, P.J. (eds) *Land Use for Agriculture, Forestry and Rural Development.* Department of Agricultural Economics and Food Marketing, University of Newcastle upon Tyne, Newcastle upon Tyne, pp. 179–193.

Mas-Colell, A., Whinston, M.D. and Green, J.R. (1995) *Microeconomic Theory.* Oxford University Press, New York.

Mathews, R.C.O. (1986) The economics of institutions and the sources of growth. *Economic Journal* 96, 903–918.

Mattsson, L. and Li, C. (1993) The non-timber value of northern Swedish forests. *Scandinavian Journal of Forest Research* 8, 426–434.

Meade, J.E. (1952) External economies and diseconomies in a competitive situation. *Economic Journal* 62, 54–67.

Merlo, M. (1996) Non-market environmental values in forest management accounting. *Finnish Journal of Business Economics* 1, 29–47.

Ministerium für Ländlichen Raum, Ernährung, Landwirtschaft und Forsten Baden-Württemberg (1993) *Richtlinie zur Förderung der Erhaltung und Pflege der Kulturlandschaft und von Erzeugungspraktiken, die der Marktentlastung Dienen (MEKA).* Az.: 64.8872.50, Stuttgart.

Ministerium für Umwelt Baden-Württemberg (1990) Landschaftspflegerichtlinie. Az.: 27 (UM)/65 (MLR)-8870.00, Stuttgart.

Mirschel, W. and Glemnitz, M. (1995) Entwicklungsmöglichkeiten für aus der landwirtschaftlichen Nutzung fallende Flächen. In: Bork, H.-R., Dalchow, C., Kächele, H., Piorr, H.-P. and Wenkel, K.O. (eds) *Agrarlandschaftswandel in Nordost-Deutschland unter veränderten Rahmenbedingungen: ökologische und ökonomische Konsequenzen.* Ernst & Sohn Verlag, Berlin, pp. 84–96.

Mitchell, R.C. and Carson, R.T. (1989) *Using Surveys to Value Public Goods: the Contingent Valuation Method.* Resources for the Future, Washington, DC.

MLNV (Ministerie van Landbouw, Natuurbeheer en Visserij) (1990) *Natuurbeleidsplan: Regeringsbeslissing.* SDU, The Hague.

MLNV (Ministerie van Landbouw, Natuurbeheer en Visserij) (1992) *Structuurschema Groene Ruimte.* SDU, The Hague.

MLNV (Ministerie van Landbouw, Natuurbeheer en Visserij) (1994) *Rijksbegroting 1995.* SDU, The Hague.

Montresor, E. (1994) Il ruolo dell'agricoltura nelle aree protette. *La Questione Agraria* 55, 135–148.

Moxey, A.P., White, B. and O'Callaghan, J.R. (1995a) The economic component of NELUP. *Journal of Environmental Planning and Management* 38, 21–34.

Moxey, A.P., White, B., Sanderson, R.A. and Rushton, S.P. (1995b) An approach to linking an ecological vegetation model to an agricultural economic model. *Journal of Agricultural Economics* 46, 381–397.

Mueller, D.C. (1989) *Public Choice II.* Cambridge University Press, Cambridge, 518 pp.

Mulder, M. and Poppe, K.J. (1993) *Landbouw, Milieu en Economie.* Periodieke Rapportage 68–89, LEI-DLO, The Hague.

Musu, I. and Siniscalco, D. (eds) (1993) *Primo Rapporto Intermedio della Commissione ISTAT-FEEM per lo Studio di un Sistema di Contabilità Ambientale, Ambiente e Contabilità Nazionale.* Il Mulino, Bologna, pp. 15–49.

Naiman, R.J. (ed.) (1992) *Watershed Management: Balancing Sustainability and Environmental Change.* Springer-Verlag, New York.

Navrud, S. (ed.) (1992) *Pricing the European Environment.* Scandinavian University Press, Oslo.

Neher, P.A. (1990) *Natural Resource Economics: Conservation and Exploitation.* Cambridge University Press, Cambridge.

Norgaard, R.B. (1988) Sustainable development: a coevolutionary view. *Futures* December, 606–620.

Norgaard, R.B. (1992a) Sustainability: the paradigmatic challenge to agricultural economists. In: Peters, G.H. and Stanton, B.F. (eds) *Proceedings of the Twenty-First Conference of Agricultural Economist.* Dartmouth Publishing Company, Aldershot, pp. 92–101.

Norgaard, R.B. (1992b) Coordinating disciplinary and organizational ways of knowing. *Agriculture, Ecosystems and Environment* 42, 205–216.

O'Callaghan, J.R. (ed.) (1996) *The Interaction of Economics, Ecology and Hydrology.* Chapman and Hall, London.

Oglethorpe, D.R. and O'Callaghan, J.R. (1995) Farm-level modelling within a river catchment decision support sytem. *Journal of Environmental Planning and Management* 38, 77–92.

Oglethorpe, D.R., Sanderson, R.A. and O'Callaghan, J.R. (1995) The economic and ecological impact at the farm level of adopting Pennine Dales Environmentally Sensitive Area grassland management prescriptions. *Journal of Environmental Planning and Management* 38, 125–136.

O'Neill, J. (1992) The varieties of intrinsic value. *The Monist* 75, 119–133.

Oskam, A.J. (1994) *Het Landbouw/Natuur-vraagstuk: Economisch Gezien.* Vakgroep Algemene Agrarische Economie, Wageningen.

Oude Lansink, A. (1994) Effects of input quota in Dutch arable farming. *Tijdschrift voor Sociaal Wetenschappelijk Onderzoek van de Landbouw* 9, 197–217.

Onde Lansink, A. and Peerlings, J. (1996) Modelling the new EU cereals and oilseeds regime in the Netherlands. *European Review of Agricultural Economics* 24, 161–178

Owen, D. (1992) *Green Reporting, Accountancy and the Challenge of the Nineties.* Chapman and Hall, London.

Parker, K. (1990) *Two Villages, Two Valleys. The Peak District Integrated Rural Development Project 1981–88.* Witley Press, Hunstanton, 116 pp.

Pearce, D.W. and Turner, R.K. (1990) *Economics of Natural Resources and the Environment.* Harvester Wheatsheaf, Exeter, 359 pp.

Peerlings, J. (1993) An applied general equilibrium model for Dutch agribusiness policy analysis. PhD Thesis, Agricultural University, Wageningen, 254 pp.

Perman, R., Ma, Y. and McGilvray, J. (1996) *Natural Resource and Environmental Economics.* Longman, London.

Perrings, C. and Walker, B.W. (1995) Biodiversity loss and the economics of discontinuous change in semiarid rangelands. In: Perrings, C., Maler, K.-G., Folke, C., and Jansson, B.-O., (eds) *Biodiversity Loss: Economic and Ecological Issues.* Cambridge University Press, Cambridge, pp.190–210.

Peskin, H.M. and Lutz, E. (1990) *A Survey of Resource and Environmental Accounting in Industrialised Countries.* Environment Working Paper 37, World Bank, Washington, DC.

Pimentel, D., Stachow, U., Takacs, D.A., Brubaker, W., Dumas, A.R., Meany, J.J., O'Neil, J.A.S., Onsi, D.E. and Corzilius, D.B. (1992) Conserving biological diversity in agricultural/forestry systems. *BioScience* 42, 354–362.

Piorr, H.-P., Dannowski, R., Deumlich, D., Glemnitz, M., Kächele, H., Kersebaum, K.C., Mirschel, W. and Stachow, U. (1997) Ecological and socioeconomic consequences of the European Union reform policy after 1996 in North-East Germany. In: Romstad, E., Simonsen, J. and Vatn, A. (eds), *Controlling Mineral Emissions in European Agriculture.* CAB International, Wallingford, pp. 261–284.

van der Ploeg, J.D. (1994) Styles of farming: an introductory note on concepts and methodology. In: Van der Ploeg, J.D. and Long, A. (eds) *Born from Within: Practice and Perspectives of Endogenous Rural Development.* Van Gorcum, Assen.

van der Ploeg, J.D. (1995) Inside the melting pot: some empirical observations on entreprise networks in dutch agriculture. In: CESAR (ed.) *Il Sistema di Agrimarketing e le Reti d'Impresa.* Università degli Studi di Perugia, Perugia, pp. 229–241.

van der Ploeg, J.D. and Saccomandi, V. (1995) On the impact of endogenous development in agriculture. In: van der Ploeg, J.D. and Van Dijk, G. (eds) *Beyond Modernization.* Van Gorcum, Assen, pp. 10–27.

Poggi, A. (1995) Una storia infinita: la protezione dell'area del delta del Po. In: Lega Ambiente dell'Emilia-Romagna (ed.) *Eco-annuario.* CDS Edizioni, Ferrara, pp. 163–173.

Porter, M. (1991) *Il Vantaggio Competitivo delle Nazioni*. Mondadori, Milan, 757 pp.

Powe, N., Garrod, G., Brunsdon, C. and Willis, K. (1995) *Estimating the Benefits of Woodland Access Using an Hedonic Price Approach Based on Data from a Geographic Information System*. Working Paper 18, CRE, University of Newcastle upon Tyne, Newcastle upon Tyne.

Pretty, J.N. (1995) *Regenerating Agriculture: Policies and Practice for Sustainability and Self-Reliance*. Earthscan, London.

Randall, A. (1993) What practicing agricultural economists really need to know about methodology. *American Journal of Agricultural Economics* (75th Anniversary Issue), 48–59.

Reiling, S.D., Boyle, K.J., Phillips, M.L. and Anderson, M.W. (1990) Temporal reliability of contingent values. *Land Economics* 66(May), 128–134.

Repetto, R., Magrath, W., Wells, M., Beer, C. and Rossini, F. (1989) *Wasting Assets: National Resources in the National Income Accounts*. World Resources Institute, Washington, DC.

Richard, A. and Trommetter, M. (1994) La rationalisation des contracts entre pouvoirs publics et agriculteurs: le cas de mesures agri-environnementales. In: INRA (ed.) *Actes et Communications, 12. Réformer la Politique Agricole Commune. L' Apport de la Recherche Économique*. INRA, Paris, pp. 307–324.

Romer, P. (1994) The origins of endogenous growth. *Journal of Economic Perspectives* 8, 3–22.

Rosen, H.S. (1995) *Public Finance*, 4th edn. Irwin, Homewood, Illinois, 623 pp.

Rosenberg, A. (1992) *Economics: Mathematical Politics or Science of Diminishing Returns?* University of Chicago Press, Chicago.

RSU (Rat der Sachverständigen für Umweltfragen) (1994) *Umweltgutachten 1994 – Für eine dauerhaft-umweltgerechte Entwicklung*. Bundesanzeiger Verlagsgesellschaft, Bonn.

Rushton, S.P., Cherril, A.J., Tucker, K., Sanderson, R.A., and O'Callaghan, J.R. (1995) The ecological modelling system of NELUP. *Journal of Environmental Planning and Management* 38, 35–52.

Ruttan, V.W. (1996) What happened to technology adoption–diffusion research? *Sociologia Ruralis* 36, 51–73.

Sagoff, M. (1988) *The Economy of the Earth*. Cambridge University Press, Cambridge, 262 pp.

Sagoff, M. (1994) Should preferences count? *Land-Economics* 70, 127–144.

Sammarco, G. (1993) Approcci alla contabilità: problemi e proposte. In: Musu, I. and Siniscalco, D. (eds) *Ambiente e Contabilità Nazionale*. Il Mulino, Bologna, pp. 65–92.

Saunders, C.M. (1996) *Financial, Public Exchequer and Social Value of Changes in Agricultural Output*. Working Paper, Centre for Rural Economy, Department of Agricultural Economics and Food Marketing, University of Newcastle upon Tyne, Newcastle upon Tyne.

Scheele, M. (1996) The agri-environmental measures in the context of the CAP reform. In: Whitby, M. (ed.) *The European Environment and CAP Reform: Policies and Prospects for Conservation*. CAB International, Wallingford, pp. 3–7.

Schkade, D.A., and Payne, J.W. (1993) Where do the numbers come from? How people respond to contingent valuation questions. In: Hausman, J.A. (ed.) *Contingent Valuation: a Critical Assessment*. North-Holland, Amsterdam, pp. 271–293.

Schumpeter, J. (1954) *History of Economic Analysis*. Allen and Unwin, London.

Sijtsma, F.J. and Strijker, D. (1995) *Effect-analyse Ecologische Hoofdstructuur, Deel I en Deel II* Stichting Ruimtelijke Economie, Groningen.

Simons, B., Burrell, A.M., Oskam, A.J., Peerlings, J.H.M. and Slangen, L.H.G. (1994) *CO_2 Emissions of Dutch Agriculture and Agribusiness: Method and Analysis.* Wageningen Economic Papers 1994–1, Wageningen.

Siniscalco, D. (ed.) (1995) *La Contabilità Ambientale d'Impresa.* Il Mulino, Bologna, 44 pp.

Slangen, L.H.G. (1994a) *De Financiele en Economische Aspecten van het Natuurbeleidsplan.* Vakgroep Algemene Agrarische Economie, Wageningen, 27 pp.

Slangen, L.H.G. (1994b) The economic aspects of environmental co-operatives for farmers. *International Journal of Social Economics* 21(9), 42–59.

Smith, V.K. (1992a) Arbitrary values, good causes, and premature verdicts. *Journal of Environmental Economics and Management* 22, 71–89.

Smith, V.K. (1992b) Environmental costing for agriculture: will it be standard fare in the Farm Bill of 2000? *American Journal of Agricultural Economics* 74, 1076–1088.

de Snoo, G.R. (1995) Unsprayed field margins: implications for environment, biodiversity and agricultural practice. PhD thesis, Leiden University.

Spaninks, F.A. (1993) *Een Schatting van de Sociale Baten van Beheersovereenkomsten met Behulp van de Contingent Valuation Methode.* Scriptie, Vakgroep Algemene Agrarische Economie, Wageningen.

Spash, C.L. and Hanley, N. (1995) Preferences, information and biodiversity preservation. *Ecological Economics* 12, 191–208.

Spasiano, C. (1992) E'un primato italiano il pool antinquinamento. *L'Impresa Ambiente* 3, 23–29.

Stachow, U. (1995) Naturraum- und Biotopausstattung. In: Bork, H.-R., Dalchow, C., Kächele, H., Piorr, H.-P. and Wenkel, K.O. (eds) *Agrarlandschaftswandel in Nordost-Deutschland unter veränderten Rahmenbedingungen: ökologische und ökonomische Konsequenzen.* Ernst & Sohn Verlag, Berlin, pp. 286–321.

Steinhauser, H., Langbehn, C. and Peters, U. (1992) *Einführung in die Landwirtschaftliche Betriebslehre 1.* Eugen Ulmer, Stuttgart.

Stern, A. (ed.) (1993) *La gestion de l'Environnement en Europe. Guide de l'Éco-audit et de l'Éco-label.* Club de Bruxelles, Brussels.

Sudit, E.F. and Whitcomb, D.K. (1976) Externality production functions. In: Lin, S.A.Y. (ed.) *Theory of Economic Externalities.* Academic Press, New York.

Sumpsi, J. (1994) La agricultura española actual. El marco de referencia. *Papeles de Economía Española* 60/61, 2–14.

Sumpsi, J. (1996) Problems of the environmental measures. *Discussion paper submitted to the Integrated Rural Policy Group.* D.G. VI, Commission of the European Communities, Brussels, unpublished.

Teisl, M.F., Boyle, K.J., McCollum, D.W. and Reiling, S.D. (1995) Test–retest reliability of contingent valuation with independent sample pretest and posttest control groups. *American Journal of Agricultural Economics* 77, 613–619.

Tempesta, T. (1994) Agricoltura e parchi: conflittualità e sinergie nell'esperienza veneta. In: Prestamburgo, M. and Tempesta, T. (eds) *Sistemi Produttivi, Redditi Agricoli e Politica Ambientale.* Angeli, Milan, pp. 38–66.

Terwan, P. and van der Bijl, G. (1996) Dutch country report. In: Umstätter, J. and Dabbert, S. (eds) *Policies for Landscape and Nature Conservation in Europe. An Inventory to Accompany the Workshop on Landscape and Nature Conservation held on 26–29 September 1996 at the University of Hohenheim, Germany.* Department of

Farm Management, Production Theory (and Resource Economics Unit), University of Hohenheim, Stuttgart.

Timmerman, J.T.J and Vijn, K.J. (1996) Modelmatige verweving landbouw und natuur. MSc thesis, Department of Farm Management, Wageningen Agricultural University, Wageningen.

Tomasin, A. (1990) *L'Ipotesi di Parco del Delta del Po: Materiali di Analisi.* Quaderni della Cassa di Risparmio di Padova e Rovigo, Padua, 351 pp.

Topp, K. (1993) *Linking Biological and Economic Models. Report of Workshop Proceedings held at the Royal Botanic Gardens, Edinburgh 26th to 27th October 1993.* Scottish Agricultural Statistics Service, University of Edinburgh, Edinburgh.

Turner, M.G. and Gardner, R.H. (eds) (1991) *Quantitative Methods in Landscape Ecology.* Springer-Verlag, New York.

Tversky, A. and Kahneman, D. (1974) Judgement under uncertainty: heuristics and biases. *Science* 185, 1124–1131.

United Nations (1993) *Handbook of National Accounting, Integrated Environmental and Economic Accounting, Interim version.* Department for Economic and Social Information and Policy Analysis, UN, New York.

US General Accounting Office (1989) *Farm Programs: Conservation Reserve Program Could be Less Costly and More Effective.* GAO/RCED-90-13, US General Accounting Office, Washington, DC.

Vadnjal, D. and O'Connor, M. (1994) What is the value of Rangitoto Island? *Environmental Values* 3, 369–380.

Van der Hamsvoort, C.P.C.M. and Latacz-Lohmann, U. (1996) *Auctions as a Mechanism for Allocating Conservation Contracts Among Farmers.* Onderzoekverslag, The Hague, 47 pp.

Van der Meulen, H.A.B., de Snoo, G.R. and Wossink, G.A.A. (1996) Farmers' perception of unsprayed crop edges in The Netherlands. *Journal of Environmental Management* 47, 241–255.

van der Weijden, W.J. and Timmerman, E.A. (1994) *Integrating the Environment with the EU Common Agricultural Policy.* Centre for Agriculture and Environment, Utrecht.

Verdiesan, F. and Moxey, A.P. (1994) *Disaggregating Land Cover Changes with a Spatial Influence Algorithm.* NELUP Technical Report No. 40, CLUWRR, University of Newcastle upon Tyne, Newcastle upon Tyne.

Victor, P.A. (1991) Indicators of sustainable development. some lessons from capital theory. *Ecological Economics* 4, 191–213.

Wallace, W.A. (ed.) (1994) *Ethics in Modelling.* Pergamon Press, Oxford.

Wallis, J.J. and North, D.C. (1986) Measuring the transaction sector in the American economy, 1870–1970. In: Engerman, S.L. and Gallman, R.E. (eds) *Long Term Factors in American Economic Growth.* University of Chicago Press, Chicago.

Walsh, R.G., Johnson, D.M. and McKean, J.R. (1992) Benefit transfer of outdoor recreation demand studies, 1968–1988. *Water Resources Research* 28(3), 707–713.

Weaver, R.D. (1996) Prosocial behavior: private contributions to agriculture's impact on the environment. *Land Economics* 72, 231–247.

Weaver, R.D. and Harper, J.K. (1993) Analysing water quality policy using microeconomic models of production practices and biophysical flow models of environmental processes. *Resource Management and Optimization* 9, 95–105.

Weaver, R.D., Harper, J.K. and Gillmeister, W.J. (1996) Efficacy of standards versus incentives for managing the environmental impacts of agriculture. *Journal of Environmental Management* 46, 173–188.

Weber, J.L. (1986) Le système de comptes du patrimoine naturel. In: INSEE (ed.) *Les Comptes du Patrimoine Naturel*. Les Collection de l'INSEE, serie C, Paris, pp. 65–126.

Weinschenck, G. (1994) Rückkehr zu den Prinzipien praktischer Vernunft. *Agrarwirtschaft* 43, 97–98.

Weinschenck, G. (1995) Zwischen Knappheit, Umweltzerstörung und Überfluß – Landwirtschaft auf dem Weg ins 21. Jahrhundert. *Agrarwirtschaft* 44, 331–335.

Whitby, M.C. (ed.) (1994) *Incentives for Countryside Management: the Case of Environmentally Sensitive Areas*. CAB International, Wallingford.

Whitby, M.C. and Saunders, C.M. (1996) Estimating the supply of conservation goods in Britain: a comparison of the financial efficiency of two policy instruments. *Land Economics* 72, 313–325.

Whitehead, J.C., and Hoban, T.J. (1996) Testing for temporal reliability in contingent valuation with time for changes in factors affecting demand. Draft.

Whittaker, J.M., O'Sullivan, P. and McInerney, J.P. (1991) An economic analysis of management agreements. In: Hanley, N. (ed.) *Farming and the Countryside: an Economic Analysis of External Costs and Benefits*. CAB International, Wallingford, pp. 196–214.

Williamson, O. (1975) *Markets and Hierarchies: Analysis and Antitrust Implications*. The Free Press, New York.

Williamson, O. (1985) *The Economic Institutions of Capitalism*. The Free Press, New York.

Willis, K. (1991) The recreational value of the Forestry Commission Estate in Great Britain: a Clawson–Knetsch travel cost analysis. *Scottish Journal of Political Economy* 38, 58–75.

Willis, K.G. and Benson, J.F. (1993) Valuing environmental assets in developed countries. In: Turner, R.K. (ed.) *Sustainable Environmental Economics and Management: Principle and Practice*. Belhaven Press, London, pp. 269–295.

Willis, K.G. and Garrod, G.D. (1993) Valuing landscape: a contingent valuation approach. *Journal of Environmental Management* 37, 1–22.

Willis, K.G. and Garrod, G.D. (1994) The ultimate test: measuring the benefits of ESAs. In: Whitby, M. (ed.) *Incentives for Countryside Management: the Case of Environmentally Sensitive Areas*. CAB International, Wallingford, pp. 179–217.

Withrington, D. and Jones, W. (1992) The enforcement of conservation legislation. In: Howarth, W. and Rodgers, C.P. (eds) *Agriculture, Conservation and Land Use: Law and Policy Issues for Rural Areas*. University of Wales Press, Cardiff, pp. 90–107.

Wolf, S.A. and Allen, T.F.H. (1995) Recasting alternative agriculture as a management model: the value of adept scaling. *Ecological Economics* 12, 5–12.

World Commission on Environment and Development (1987) *Our Common Future*. Oxford University Press, Oxford.

Wossink, G.A.A. and Jurgens, C.R. (1995) Strategic planning for sustainable land use: prospects from rural development and farm economic. In: Sotte, F. and Zanoli, R. (eds) *The Regional Dimension of Agricultural Economics and Politics*. Ancona, pp. 631–642.

Wossink, G.A.A., Buys, J.C., Jurgens, C.R., de Snoo, G.R. and Renkema, J.A. (1996) What, how and where: nature conservation and restoration in sustainable agriculture. Paper contributed to the VIIth European Congress of Agricultural Economists, Edinburgh, 3–7 September.

Zeddies, J. and Doluschitz, R. (1996) *Marktentlastungs- und Kulturlandschaftsausgleich (MEKA). Wissenschaftliche Begleituntersuchung zu Durchführung und Auswirkungen.* Agrarforschung in Baden-Württemberg, Vol. 25, Ulmer, Stuttgart, 323 pp.

Zey, M. (ed.) (1992) *Decision Making: Alternatives to Rational Choice Model.* Sage Publications, London.

Index